HOW DO YOU GET AN EGG INTO A BOTTLE?

THIS IS A CARLTON BOOK

First published in Great Britain in 2010 by Carlton Books Limited
an imprint of the Carlton Publishing Group
20 Mortimer Street
London W1T 3JW

10 9 8 7 6 5 4 3 2

A catalogue record for this book is available from the British Library

ISBN 978-1-84732-527-3

Printed in Great Britain by CPI MacKays, Chatham, ME5 8TD

Authors' acknowledgments
The laws of nature are immutable. They are, however, not recorded in any
statute books that might be available to humanity to study the rules that govern
our cosmos. So, in a laborious process going back to the dawn of history,
scientific man has tried to fathom the mysteries of the universe. These collective
efforts, spanning millennia, provide a rich source of challenging conundrums.
In compiling this book, we enjoyed permission from other authors to use their
material, including Scot Morris, Stephen Barr, and the *Encyclopedia Britannica*.
Thanks should go to Sarah Gerrard for her patient support, and to Bob
Anthony on whom we tested some of our ideas.

HOW DO YOU GET AN

EGG INTO A BOTTLE?

AND OTHER QUESTIONS

ERWIN BRECHER
AND MIKE GERRARD

CARLTON
BOOKS

Contents

Introduction

by Erwin Brecher

"*All men by nature desire knowledge*" —Aristotle (384–322 BC)

This book is for the curious.

We are surrounded by phenomena often dressed in everyday clothes which are rarely recognized as such until, at odd moments, we stop to ask ourselves: What arcane rules make the physical world behave as it does?

Whereas philosophy is concerned with the "why" in probing the innermost corners of spiritual existence, science explores the "how" by processes of inductive and deductive reasoning. The former process is an exercise in arriving at general hypotheses and theories from specific observations and experiments, while the latter reverses the procedure by starting with the theory to show that it will satisfactorily explain experimental results. This dichotomous approach is at the core of the scientific method, which was first articulated during the Renaissance but which in fact has been used throughout human history.

More often than not we take the workings of the physical world around us for granted without realizing the profound effect these, many apparently mundane, phenomena have on our understanding of the universe.

Sir Isaac Newton (1643–1727), one of the great scientists of all time, discovered the principles of universal gravitation following the fall of an apple. The story might be anecdotal, but Newton's achievements are monumental…

The time is New Year's Eve 1670. Picture yourself sitting at home and seeing an object fall. Would you have given it a second thought? Probably not. But if you had, you would most likely have said to yourself: "What else could it do but fall to the floor?" In other words, you would have considered it axiomatic that everything falls down and not up. However there are no axioms in physics; that is to say, there are no basic principles that are assumed to be true without proof or inherent logic, as is frequently the case in geometry.

On the contrary, there is not a single aspect of our physical world which is not investigated, analysed and evaluated by scientists who are determined to find an explanation for all phenomena, be they esoteric or mundane. The relevant literature is enormous, with some questions remaining unresolved while opinions on other theories differ.

We can't all be Newtons, but we can have an inquiring mind and derive a great deal of satisfaction from understanding what makes things "tick." In some cases we shall find the answer, using basic principles of physics and a modicum of problem-solving ability. With others the answer will make immediate sense, with a tinge of regret for not having thought of the explanation in the first place. Finally there will be quite a few where the solution will be a revelation.

In all cases, the reader will close this book feeling that his horizon has widened beyond expectation.

Erwin Brecher

Introduction
by Mike Gerrard

I have always loved puzzles of all sorts, from crosswords to detective novels. There is something immensely satisfying about tracking down the solution from a few apparently unconnected bits of information. I think that is why I have always been fascinated by science. The scientific method of proposing hypotheses and then checking them out by experimentation is immensely attractive.

But solving a scientific conundrum has an additional attraction that no thriller writer can hope to match. Having worked out that "the butler must have done it" is satisfying, but at the end of the day it is just a story. Solving a scientific puzzle also teaches you something about the way the universe works.

It must be one of the most exciting and creative of human activities to discover something absolutely new that no one has ever known before. The Einsteins, Newtons, Darwins, and so on, must all have shared an incredible sense of elation having made that final connection in their minds.

This book does not pretend to be in quite the same league, but it is hoped that you can experience a similar sensation. You may find an observation that you have never seen for yourself, or a question that you have never thought to ask.

The solution to it lies in your own head somewhere; all you need do is formulate your own thoughts to produce the solution. The process can be completely absorbing; it is where the stories of absent-minded professors originate. Archimedes was so involved with his own problem that he foolishly asked an invading Roman soldier to get out of the light, and got himself killed.

To get the most out of this book you are advised not to rush between the problems and the stated solutions; that will spoil the fun! Think about things. It may take several days or a little research to decide on what you think the answer is. Only then look at our solution. You may even disagree with what we say – your solution may be superior to ours. Happy solving!

Mike Gerrard

How do I blow up my stereo?

When setting up my new hi-fi system, I noticed that the instructions were very emphatic about connecting the wires to both speakers in the same way. In fact, the wires were colour-coded to try to make it difficult to get wrong.

What would happen if the loudspeakers were incorrectly connected?

Solution on page 115

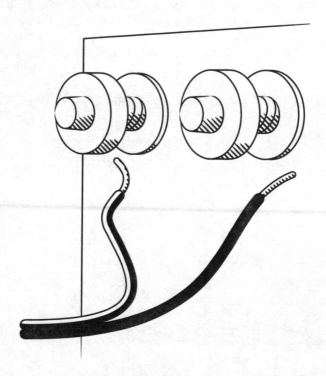

How can I turn red to green?

A light-emitting diode (LED) is a small solid-state device that gives off light when a current is passed through it. It is commonly used in clocks to give a clear red, or sometimes green, digital display.

There is a type of LED on the market that glows red when the current is passed one way through it, and green when the current flows the other way. An alternating current flows rapidly one way and then the other. What would this LED look like if it were connected to an alternating source?

Solution on page 115

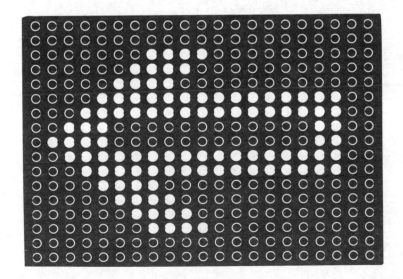

Why does the ringing change?

I was recently putting in some fencing in a field. While knocking in the fence posts, I noticed differences in the sound. In the centre of the field I was a long way from anything, and the sound from the hammering of the posts was dull and flat. I concluded that this was because there was nothing near to reflect the sound. In another part of the field I could hear a distinct echo from a nearby building.

However, in a third part of the field I could hear a distinct ringing tone. Can you think what might have been the cause of the ringing sound?

Solution on page 115

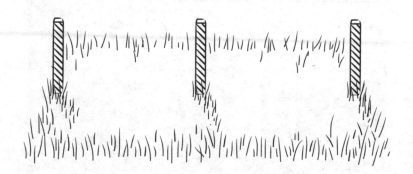

Why can't I tune my radio?

On a recent drive around the country, I was listening to a very interesting program on the car radio. As I moved away from a transmitter, the reception got progressively worse, so I had to retune the radio to pick up the same station on a nearer transmitter. I had to repeat this process several times during the journey. This really affected my enjoyment of the program. Why do they not ensure that all transmitters broadcasting the same station use the same frequency?

Solution on page 116

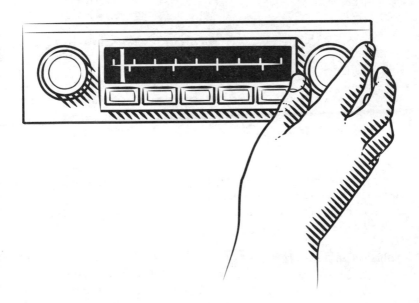

Are there rainbows on the Moon?

Rainbows, an enchanting spectacle of nature, pose a number of baffling questions. We know they are caused by drops of water falling through the air, refracting sunlight in such a way that it creates a curve of light exhibiting all colours of the spectrum in their natural order. Now answer the following:

1. Why is it you do not always see a rainbow when it rains while the Sun shines?
2. Can the Moon also produce a rainbow?
3. Even without rain you can, at times, see a rainbow if you look across a lawn early in the morning. Why?

Solution on page 116

How do I dull my headlights?

Many night driving accidents are caused by the glare of headlights from approaching cars. My friend Jonathan, who fancies himself an inventor, came up with an idea offering an effective and inexpensive solution to the problem.

"Let us introduce polarizing filters in front of headlights," he said, "to polarize light horizontally, which would absorb all photons whose electric vectors are vertical. Conversely, let us use windshields with a filter turned 90° to the first, absorbing light emitted by the headlights." On the face of it, this would be the perfect solution, as the light from the approaching car would be blocked out, while all other objects would be visible.

Would this idea work? If so, why hasn't it been adopted?

Solution on page 116

How do I make a toy boat?

When, as a boy, I built my own little steamboat to operate in the bathtub, I was thrilled. It was my own design, using odd bits of metal plus a little fatherly help, and looked something like this:

The principle was simple. The candle heated the water in the boiler and the resulting steam forced the water through the outlets as jets, propelling the boat forward. To my surprise I noticed that, from time to time, the outward jets stopped and instead water was sucked in through the tubes. I did not understand why this should be and, more surprising, why the boat did not then reverse direction and move backwards, as a basic law of physics would suggest (Newton's Third Law of Motion: action = reaction).

Can you explain?

Solution on page 116

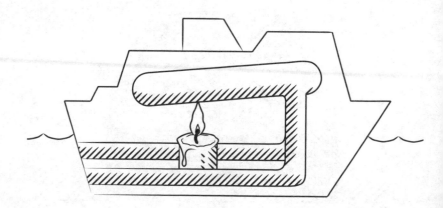

Does wind affect my mind?

It has long been accepted that weather has a profound effect on the physical well-being and even the mood of most people. One of the well-known natural phenomena is a warm dry wind blowing down from high mountains into valleys. In German-speaking countries this wind is called *foehn*, though it has many other names with one thing in common: dry, warm and unpleasant. *Foehn* or its equivalent is held responsible for severe headaches and, in the extreme, for criminal behaviour in some of those affected by it.

How can a warm wind come down from a cold mountain, reach speeds of close to 130 kilometres per hour (75 miles per hour), and have such an impact on many people?

Solution on page 116

Why is the safety lamp safe?

During the 18th century, miners were at extreme risk when exposed to explosive gases. It was in 1815 that Sir Humphrey Davy (1778–1829) invented a safety lamp to be used in coal mines. A fine metal screen in the form of a cylinder covered the open flame of the miner's oil lamp.

Did the screening prevent explosive gases from entering the safety lamp, or is there a different explanation for the lamp's effectiveness?

Solution on page 117

Can a mirror turn me upside-down?

It is often said that mirrors reverse left-right but not up-down. Can you think how a single plane mirror can:

1. Reverse up-down as well?
2. Reverse up-left and down-right?
3. Reverse up-down but not left-right?

Solution on page 117

Can you square a circle?

Prolonged weightlessness can have unfortunate physiological effects on the human body and may prove a great problem on long space flights. It is well known that spinning the spacecraft can produce an effect very similar to gravity, so we have seen science fiction films where the astronauts live in circular living quarters which are rotating to provide apparently normal gravity.

However, as humans we are not used to living in round rooms. What would the effect be if the space explorers were put into a normal rectangular room rotating about a line drawn across the middle of the ceiling?

Solution on page 117

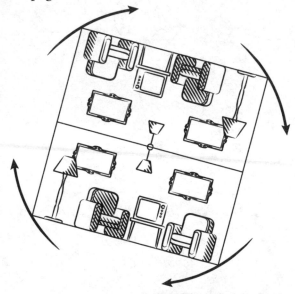

Why does the Sun go out?

Eclipses of the Sun occur when the Moon moves in front of the Sun. The Moon takes about four weeks to orbit the Earth. However, very few people have actually experienced a total eclipse of the Sun.

Why is this? Why is there not a solar eclipse monthly?

Solution on page 117

Why aren't
I floating?

An astronaut awakes in an enclosed room where gravity is apparently normal. He realizes that there are three possibilities:

1. He is subjected to gravity.
2. The spacecraft is accelerating.
3. The spacecraft is creating its own "artificial gravity" by spinning.

Is there any experiment that the astronaut can do inside the room to discover which is the true situation?

Solution on page 118

Pull of the Sun

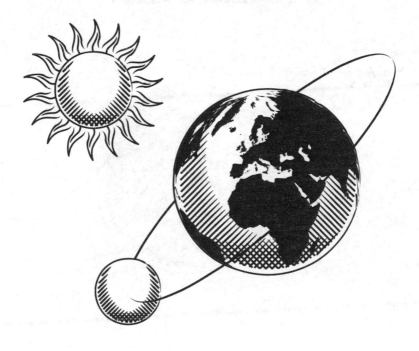

Surely the Sun's gravity pull on the Moon is much larger than the Earth's. Why then does the Sun not pull the Moon away from its orbit around the Earth?

Solution on page 118

Why do I fly back to Earth?

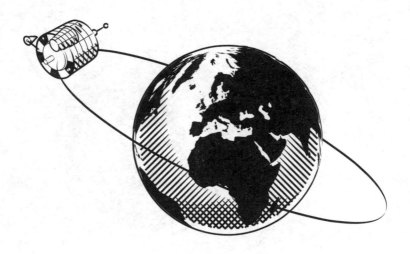

Most man-made satellites will eventually return to Earth, because their orbit is affected by the Earth's atmosphere which, though extremely thin, still exists very high up. However, the surprising effect of this air-drag is not, as one would expect, to slow the satellite down. On the contrary it will accelerate in its orbit. How do you explain this phenomenon?

Solution on page 118

What colour is the Sun?

This might seem incredibly easy to answer, but how many different answers can you think of to the question?

Solution on page 118

Why isn't the whole sky as bright as the sunlight?

Heinrich Wilhelm Matthaeus Olbers (1758–1840), a German astronomer, hypothesized that if the universe were infinite and contained an infinite number of stars, then the whole sky should be as bright as sunlight. After all, the observable universe alone indicates that the Milky Way is only one of several hundred thousand million galaxies, with each galaxy containing in turn some hundred thousand million stars. It seems therefore inconceivable that any line of sight could miss a star.

You can compare it with a blackboard on which you are asked to make a mere one million chalk marks. Surely the blackboard would appear white? Why then do we not see the sky brightly lit at night?

Solution on page 119

Why don't I burn my head?

Having washed my hair the other night, I was trying to dry it as quickly as possible. I therefore had my hair drier on its highest setting. I noticed that the air still felt very cold on my head, but burned my ears. Why?

Solution on page 119

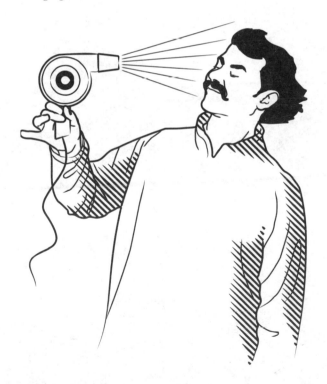

So near and yet so far?

We are very used to the idea of telescopes and binoculars, but what are they doing? Do they make distant objects look closer or bigger?

What about a magnifying glass: does it make things look bigger, or is it doing something else?

Solution on page 119

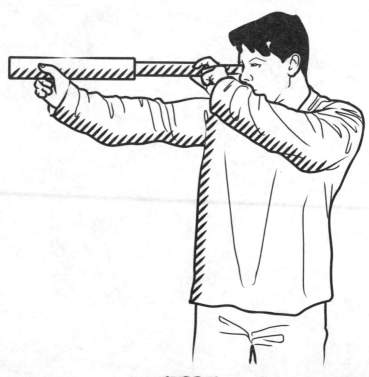

Glaringly obvious?

Ordinary sunglasses work by absorbing some of the light that would otherwise be transmitted. Polarizing sunglasses work in a slightly different way: they will transmit only light that is vibrating vertically. As light reflected from wet roads, water, and so on tends to vibrate horizontally, this cuts down glare from such surfaces. If you are wearing a pair of polarizing sunglasses and hold up another polarized pair rotated through 90 degrees, the lenses appear black because no light can now pass through both sets of glasses.

Could you insert a third set of polarizing sunglasses between the first two in such a way as to permit some light to be transmitted?

Solution on page 119

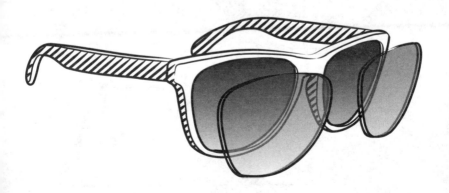

How can I balance my scales?

My kitchen scales are balanced when there is nothing on them, but I know that there is something wrong because I get different results depending on which side I put the object to be weighed. If an object balances with 100 grams (3.5 oz) when on one side, but with 144 grams (5 oz) on the other side, is it possible to determine the real weight?

Solution on page 120

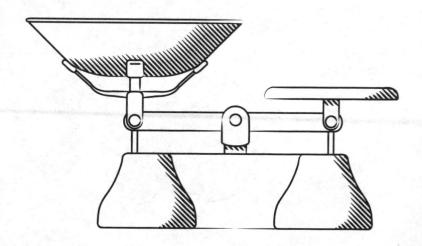

How can I tell if an egg is raw?

You have an egg before you. How can you find out whether it is hard-boiled or raw—without, of course, opening it?

Solution on page 120

Why does a golf ball have dimples?

In the early days of golf the balls were smooth. Dimples were introduced later, after manufacturers claimed the dimpled variety travelled farther. Were they right, and if so why?

Solution on page 120

How do you cool off?

You are in a hermetically sealed, perfectly heat-insulated very warm room containing a large refrigerator. Can you, more or less permanently, reduce the room temperature?

Solution on page 121

How can I make a spray gun?

As children, it was fun to annoy grown-ups by spraying them with water, using a gadget as illustrated.

By blowing through the horizontal tube, you are forcing water up the conduit tube into a fine spray. The same principle is used in aerosols and paint sprayers. Why does water rise in the tube, against gravity?

Solution on page 121

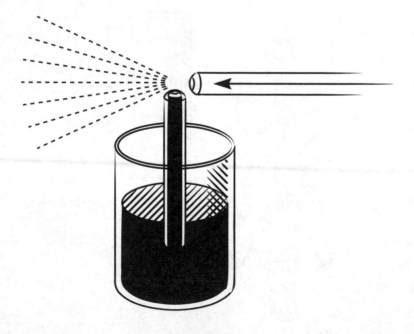

What makes glue stick?

You will probably say that this is easy to understand, but difficult to explain. If pressed further you might ascribe it to some chemical property of the adhesive. This is just begging the question, and furthermore is wrong. The answer is rather more complex. Have another guess.

Solution on page 121

What's that strip?

You will often see trucks and cars with a metal strip or chain hanging from the rear, being dragged along the ground. What is the purpose of this, and is it effective in what it is supposed to achieve?

Solution on page 121

How do you drive on ice?

Whether you are a good or bad driver makes all the difference in hazardous road conditions. Assume it is freezing and the road is covered with a layer of ice. Can you answer the following questions?

1. If you want to start the car and all you have to help you is one blanket, will you use it as underlay for the front or rear tyres?
2. Do you put the car in low or high gear? And will you quickly accelerate or maintain low speed?
3. If you succeed in moving but the car starts to skid, will you steer into the skid or maintain direction?

Solution on page 122

How do you float a balloon?

A hydrogen balloon will rise into the air. Suppose that we wanted a balloon that neither rose nor sank, but hovered at a constant height. We will do this by attaching a length of string so its weight is just enough to balance the upward force. You could do this by starting with a piece of string which was too long and successively snipping bits off until it just balanced. The trouble is that if you cut off too much, you will have to start all over again. Is there a simple way to achieve the desired effect?

Solution on page 122

When the leaves fall

In temperate zones, deciduous trees shed their leaves in the autumn. The trigger for this appears to be the shortening length of the days, even if the weather is still quite mild.

Surely it would be in the trees' interest to keep their leaves until the very last moment, in other words until the temperature fell to a dangerously low level. Can you find an explanation, considering that "Mother Nature" is an efficient professional?

Solution on page 122

Why don't the clouds move?

I was on a hiking holiday, when I noticed a peculiar phenomenon. At first, I could not work out what it was that "did not feel right." I was walking across a large plain. To the west was a range of mountains that stretched north to south as far as I could see. There was a strong wind blowing from the direction of the mountains.

I suddenly realized what seemed odd. It was the clouds. They were an unusual shape, forming long straight bands parallel to the mountains. But the really peculiar thing was that, despite the strong wind, they did not appear to be moving at all. Can you explain?

Solution on page 122

How do I bisect an object?

Suppose we have a number of flat objects, each with a different shape: square, circle, and so on. If we select any of those objects we could draw a bisector across it. I will define a bisector as a straight line that divides the object into two halves. Of course for any object we could draw an infinite number of bisectors. Would all the bisectors necessarily go through the same point for any specific object?

Solution on page 123

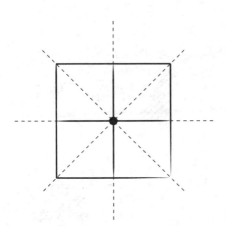

Why can sheep still eat grass?

There are many examples of two different species in an evolutionary race: for instance, cheetahs and gazelles each increasing their speed so that one can eat while the other can escape.

One might assume that similar competition exists between grass and grazing animals. Over the millennia, why has grass not evolved to be indigestible or poisonous to sheep, cows, rabbits, etc.?

Solution on page 123

How do you walk on a tightrope?

No circus act is complete without a tightrope walker. How does he keep his balance and why does he use a long bar?

Solution on page 123

Should I run through the rain?

While out walking the other day I got caught in a sudden rain shower and made a dash for home. Just as I got to my front door the rain stopped as suddenly as it had started.

While I dried out, I wondered if I would have been just as wet if I had not bothered to run. What do you think?

Solution on page 123

How do I stop my tyres wearing out?

A friend of mine lives due south of London and works due north. The M25 is the highway circling London. Every day my friend travelled up on the M25 to work by going to the east of London, and in the evening, for a change of scenery, he would travel back to the west. However, he realized that by doing this the same two tyres were on the outside on both the outward and return journeys and must have travelled farther. They must have been wearing faster. He changed his behaviour by travelling both to and from work by the easterly route.

Thinking about the situation in more detail, he realized that the lanes of the highway in one direction were outside of the lanes travelling in the other direction. So the differential wear on the tyres, although less than before, would still exist. Was my friend correct in all his reasoning?

Solution on page 124

Why don't planes fly lower?

We know that as the radius of a circle increases, so does its circumference. This must mean that the higher an airplane flies, the greater the distance it has to fly to its destination.

On a flight between London and New York, an airliner has to fly an extra 3.2 kilometres (2 miles) because of its high altitude.

Forgetting air-space congestion as a possible argument, why do airliners not fly by a shorter, lower route?

Solution on page 124

How can I sail the world?

A long time ago two captains of sailing ships had a wager with each other as to which was the faster way to sail around the world: east to west or west to east. They decided to put it to the test by having a race. So one day at exactly the same time, they set sail from a small island each going in the opposite direction.

Some months later they happened to arrive back at the same island at exactly the same time. They were just about to decide that the bet was off when they compared logs and found that there was a discrepancy of two days between them.

Is it possible to account for the discrepancy and decide who circumnavigated the world faster?

Solution on page 124

Does the Earth move?

Seismographs are sensitive instruments for detecting earthquakes. They consist of a large mass suspended by springs. When an earthquake happens, the disturbance travels through the Earth, making the instrument vibrate. However, the mass tends not to move because of its inertia. This difference in movement is amplified and written out as a trace.

Can you explain why most places in the world receive two traces for each single earthquake event? Can you also explain why a few places detect only one?

Solution on page 124

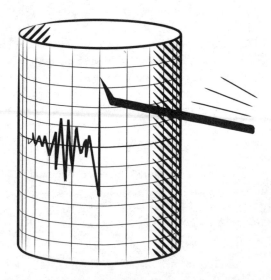

How do you burst a barrel?

There used to be a trick popular in Victorian times. A beer barrel was completely filled with water. A long thin tube had been attached to the top of the barrel, which was empty at the start of the trick. By pouring a very small amount of water into this tube the barrel could be made to burst, demonstrating that small causes can have large dramatic effects.

Suppose one jug of water was needed for the trick. If a narrower tube with half the cross-sectional area had been used, how much water would have been needed?

Solution on page 125

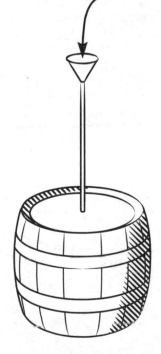

Why doesn't the snow fall evenly?

Have you noticed how much more snow is proportionally deposited on the sides of posts and poles rather than on the sides of buildings? Why?

Solution on page 125

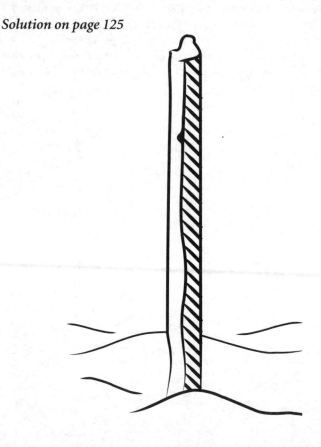

How do I shoot in a straight line?

You are standing on top of a high-rise building in Greenwich with a long-range, high-powered rifle, perfectly aligned. You are trying to hit a telegraph pole in Louth, Lincolnshire. Assume that you can see the target through the telescopic sight on your rifle. Would you aim straight at the target, to the left, or to the right of it?

To give you a clue, both Greenwich and Louth are on the 0° meridian.

Solution on page 125

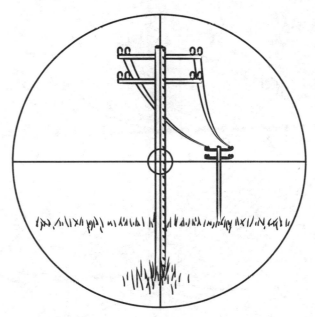

How do you drink on ice?

There are many stories of shipwrecked sailors dying of thirst or being driven insane by drinking seawater. Eskimos have no source of freshwater. Does polar ice contain salt, and if so how are the Eskimos dealing with the problem?

Solution on page 125

Why do boomerangs come back?

The boomerang is a missile used mainly by Australian aborigines. Originally used as a weapon, tossing the boomerang has recently become a sport. It is generally held vertically in the right hand, although some are designed for left-hand use. The fascinating aspect of this weapon is its ability to return to the hands of the thrower.

Can you explain this phenomenon?

Solution on page 126

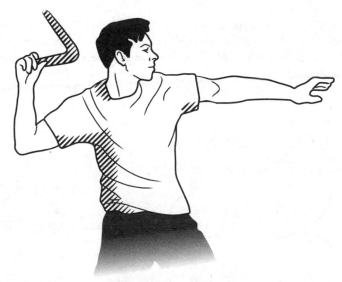

How did we avoid rockets?

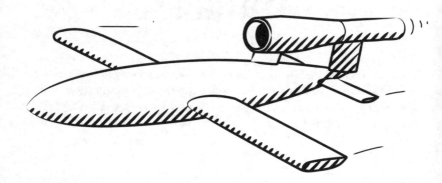

Those of us in Great Britain who lived through World War II will remember, without much nostalgia, V-1 and V-2 rockets replacing the bombing raids that had become too expensive for the Germans in terms of losses in men and planes.

It is in man's nature to be adaptable. We soon found that the V-1s presented no danger as long as you could hear them. However, as soon as the buzzing of the engine stopped, it was advisable to dive under the bed. No such strategy was available with the V-2. Why not?

Solution on page 126

How do you steer a boat?

Can you steer a boat on a lake if it is entirely becalmed and you have no oars? The answer must be no!

Suppose the circumstances are the same except that your boat is drifting in a fast-flowing river. Can you then steer with the rudder?

Solution on page 126

Can you fly faster than sound?

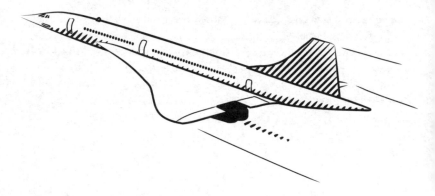

Even if you never flew in Concorde, that ingenious product of Anglo-French technology, you probably know that this plane was the only supersonic aircraft in commercial use. You may also know that Mach 1 means the speed of sound.

Now answer the following questions:

1. Do you hear the sonic boom when, travelling in Concorde, you reach Mach 1?
2. Is Concorde's ground speed always the same when the plane breaks the sound barrier?

Solution on page 126

Why is Death Valley so hot?

Death Valley is a depressed desert region in southeastern California. Most of the valley is below sea level and it has the distinction of being the hottest place in the world. In 1913 a temperature of 134°F was recorded—the highest ever at the time.

Elementary physics has taught us that hot air rises and cold air sinks. Would you therefore not have expected Death Valley to be a cool or moderately warm place, particularly as it is almost entirely enclosed by mountain ranges?

Solution on page 127

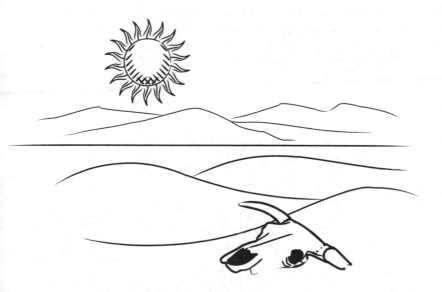

The Panama Canal conundrum

The construction of the Panama Canal is regarded as one of the greatest technical achievements of all time. It was completed ahead of schedule and was in full operation by the summer of 1914. With the map of the world in mind, one would have no doubt that the Canal runs from west to east. Surprisingly, the Pacific end lies somewhat east of the Atlantic end.

There are some other interesting aspects to which the reader is invited to find an answer:

1. At the last lock, as the gate is opened, any ship will move out to sea without tugs and without using its own power. What makes it move?
2. One would assume that the water levels in the Atlantic and Pacific are the same. However, there is a difference at times of as much as 30 centimetres (12 inches). Why are the ocean levels not the same?

Solution on page 127

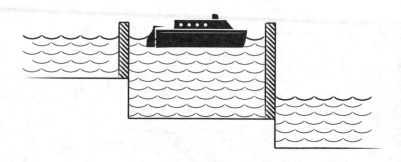

Balloon in a bath

An air-filled balloon is held underwater by a weight so that it is just on the verge of sinking; the top surface of the balloon just touches the waterline. If you push the balloon beneath the surface of the water (as shown), what will it do – bob back to the surface, remain at the level to which it is pushed, or sink to the bottom?

Solution on page 127

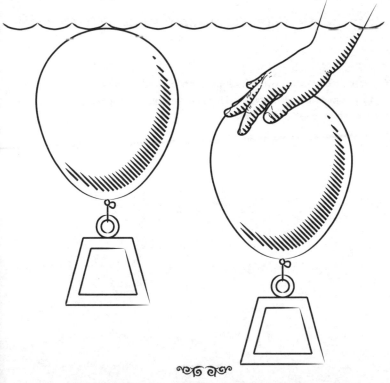

Why don't submarines rest?

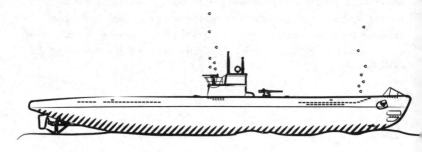

The captain of a submarine tries at all costs to avoid letting his sub come to rest on a clay or sandy ocean bottom. He knows that if this happens, it can be fatal. Why?

Solution on page 127

Does a bootstrap elevator work?

Study the drawing carefully. Can the man lift both himself and the block off the ground?

Solution on page 127

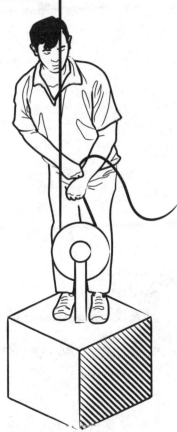

Does the balloon move?

A child sits in the back seat of the family car, holding a helium balloon on a string. All the windows are closed. As the car accelerates forward, does the balloon tilt forward, tilt backwards, or remain in the same place?

Solution on page 128

Can you make a boat on land?

Tourists on a cruise have experienced the less than enjoyable sensation of pitching and rolling in a rough sea. Sailors must either get used to it or look for another job.

On dry land, a similar sensation can be created by putting oval wheels on a vehicle. Such wheels would make it pitch backwards and forwards. Is it possible to make the vehicle also "roll," and if so, how?

Solution on page 128

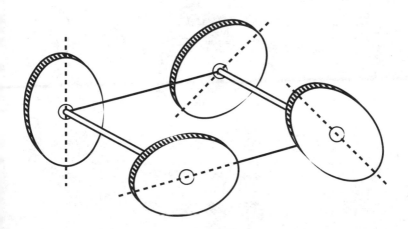

What makes a mirror reflect?

Why does a mirror reverse only the left and right sides but not up and down? Alter all, if we speak of a plain mirror, one with a surface that is perfectly smooth and flat, its left and right sides do not differ in any way from its top and bottom surfaces. Why then this persistent preference for changing left to right, while ignoring top and bottom? Can you explain this phenomenon?

While this question is indeed puzzling, we know intuitively that it could not be different. Remember, we are discussing the ordinary mirror, as it is quite possible to construct mirrors that will not reverse left and right, as well as others that will reverse top and bottom.

Solution on page 128

The tall mast

Joan and Jean are out sailing in a small boat when Jean remembers an important appointment. Joan suggests that they take a short-cut back to the dock to save time. On the way, they come to a footbridge and find that the boat's mast is slightly taller than the height of the bridge. How are they able to quickly pass under the bridge so that Jean can make her appointment? (They cannot lower the mast.)

Solution on page 128

The spectrum

One of the first art lessons learned at school is that the colour green can be obtained by mixing yellow and blue paint. On the other hand, strangely enough, if you project yellow and blue light, through appropriate gelatin sheets, onto a screen, you obtain white light.

Can you think of an explanation?

Solution on page 128

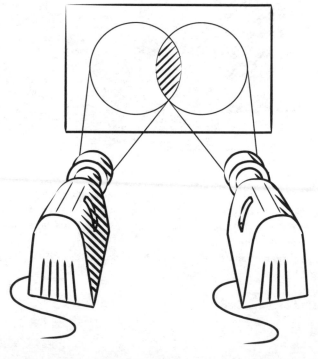

How do you open a glass jar?

Individuals using perfume bottles with glass stoppers are often confronted with stoppers that cannot be easily removed without using force that could result in breakage. Placing it under a hot water tap won't work. Yet there is an easy method that works without fail. How would you solve this problem?

Solution on page 129

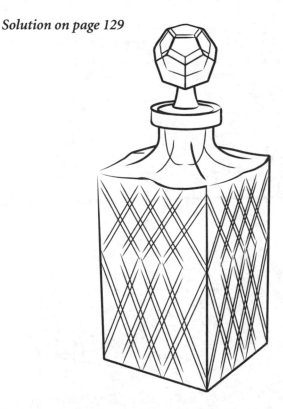

How do you make a truck fly?

There is a well-known, probably apocryphal story: A truck full of live poultry is stopped before a bridge on a country road. Its driver, beating the side of the truck with a stick, is asked what he is doing and he explains that his load is too heavy for the bridge so he is making the birds fly to lighten his load before proceeding.

This suggests the following puzzle. A cage with a bird in it, perched on a swing, weighs four pounds. Is the weight of the cage less if the bird is flying about the cage instead of sitting on the swing? Ignoring the fact that if left in an airtight box for long the bird would die, would the answer be different if an airtight box were substituted for the cage?

Solution on page 129

Solution on page 129

Hole through the Earth

Assume that a hole is drilled from one point on the globe through the centre of the Earth to the Antipodes, as illustrated, and a steel ball is dropped into the hole at point A.

Ignoring any external influences such as air resistance, friction and conditions of the Earth's core, answer the following questions:

1. As the ball is travelling from A to the Earth's centre, does its velocity increase, decrease, or stay the same?
2. Will the ball weigh less or more when it reaches the centre of the Earth?
3. Will the ball's mass change during its journey?
4. At what point will the ball be in a state of zero gravity?
5. If the ball fell through such a cylindrical hole through the centre of the Moon, would the one-way journey take more or less time than it would on Earth?

Solution on page 129

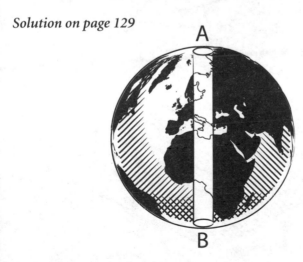

Feathers and Gold II

Again which weighs more, an ounce of feathers or an ounce of gold?

Solution on page 130

The half-hidden balance

The horizontal line is a weightless rod, balanced on a fulcrum at F, and of unknown length to the right of the fulcrum. The visible part is 1 foot long, and supports a 454-gramme (1-pound) weight. Besides being weightless, the rod is able to support any weight, and, since it is in balance, there must be a weight to the right. The further the second weight is to the right, the less it can be. What are the lower and upper limits of the total possible force downwards at F?

Solution on page 130

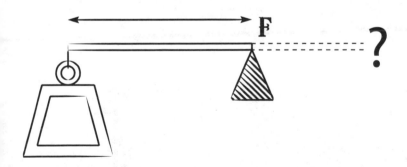

How do you make
salad dressing?

Jill and Joe went out for a picnic. "I hope you remembered to bring the oil and vinegar for the salad," said Joe.

"I certainly did," replied Jill, "and to save myself having to carry two bottles I put both the oil and vinegar in the same bottle"

"That wasn't very clever," said Joe, "because, as you very well know, I like a lot of oil and very little vinegar, but you like a lot of vinegar and hardly any oil."

Jill sighed, and then proceeded to pour, from the single bottle, exactly the right proportions of oil and vinegar that each of them wanted. How did she do it?

Solution on page 130

Hourglass puzzle

You can sometimes find an unusual toy in the shops: it is a glass cylinder full of water, with a sand-filled hourglass floating at the top.

When the cylinder is turned upside down, as in the right-hand drawing, something rather strange happens. The hourglass stays at the bottom of the cylinder until a certain amount of sand has flowed from its upper compartment into its lower compartment, then it rises slowly to the top. Can you suggest a simple explanation for this phenomenon?

Solution on page 130

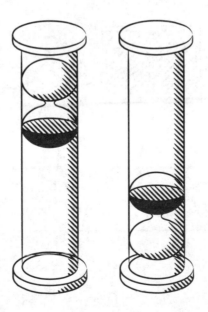

Can you blow up a room?

You are in a room filled with 100% methane gas. What would happen if you struck a match?

Solution on page 131

Rotation counter

If a bicycle with equal wheels has a rotation counter on each, why will the front one tend to give a higher reading?

Solution on page 131

Can you propel a boat with a rope?

Assume that a rope has been tied to the stern of a small boat floating in still water. Is it possible for someone standing in the boat to propel it forward by jerking on the free end of the rope?

Now consider a space capsule drifting in interplanetary space. Could it be propelled by similar means?

Solution on page 131

How many times can you tear a card?

It is impossible in reality, but assuming that one could tear a playing card in two, put the halves together and tear again (getting four), put them together again and tear, and so on 52 times.

Do you think that the pile would be more or less than 10 miles high?

Solution on page 131

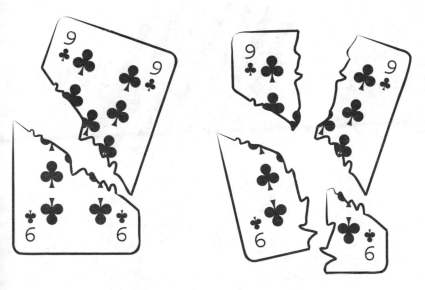

The Light Fantastic

Imagine two blacked-out rooms bathed in yellow light. A piece of white paper examined in both rooms appears to be exactly the same shade of yellow. Another piece of paper appears black in one of the rooms and striped red and green in the other. How could this be?

Solution on page 131

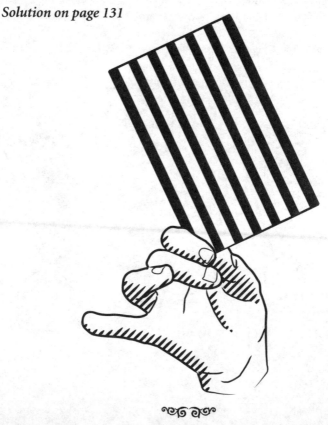

Why do only white horses survive?

One of the great intellectual achievements of all time is Darwin's theory of natural selection. There cannot be many adults in the civilized world who have not heard of Charles Darwin (1809–1882) or his *On the Origin of Species* and the phrase "survival of the fittest." We understand why polar bears are white and why grasshoppers are green. Some of us even know that animals will survive if they are able to adapt to their environment. This adaptation is only possible by random genetic mutation or by favouring those members of a species whose characteristics make them better able to cope with the hazards of the environment.

A remarkable example of the workings of Darwin's theory is the story of the horses of the Camargue—the area of the Rhone delta in France. Originally, the horses roamed the area in multicoloured herds, but eventually only the white horses survived. Can you explain this strange phenomenon?

Solution on page 132

How does a helicopter fly?

We are all more or less familiar with this heavier-than-air aircraft, which has one or more power-driven horizontal propellers that enable the craft to take off and land vertically, move in any direction, or hover stationary in the air. However, you may not have noticed the small rotor at the helicopter's tail. Is it important? What do you suppose is its purpose?

Solution on page 132

The beaker

In the diagram below, there is a cylindrical iron bar, one square centimetre in cross-section, suspended vertically over a beaker. The beaker is two square centimetres in cross-section and partly filled with water. The bar just touches the surface of the water. The beaker is standing on the right-hand pan of a balance scale. The left-hand pan contains an empty beaker and sufficient gram weights to balance the scale. One cubic centimetre of water is added to the right-hand beaker. How much water must be added to the left-hand beaker to bring the scale back to balance?

Solution on page 132

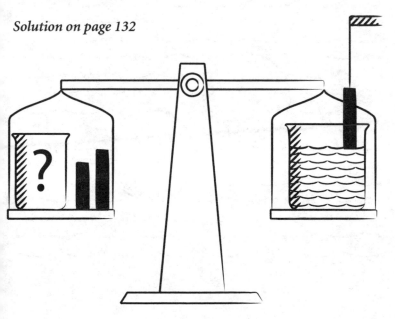

Drops and bubbles

If all space were empty except for two drops of water, the drops would be attracted to each other, according to Newton's Law of Gravity.

Now suppose all space were full of water except for two bubbles. Would the bubbles move apart, towards each other, or not at all?

Solution on page 132

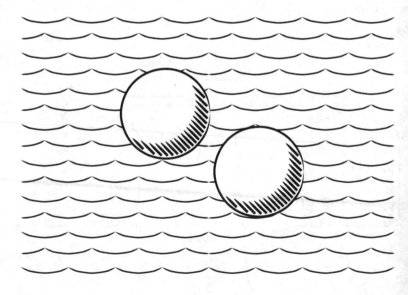

Why can't you copy a banknote?

It is illegal to make a photograph of a banknote, even if there is no intention of passing it, yet it is allowed on television. Why?

Solution on page 132

Rope and pulley

The diagram shows a rope passed over a frictionless pulley. To one end a weight is suspended, which exactly balances a man at the other end. What happens to the weight if the man attempts to climb the rope? Assume the rope to be weightless and the wheel to be frictionless.

Solution on page 132

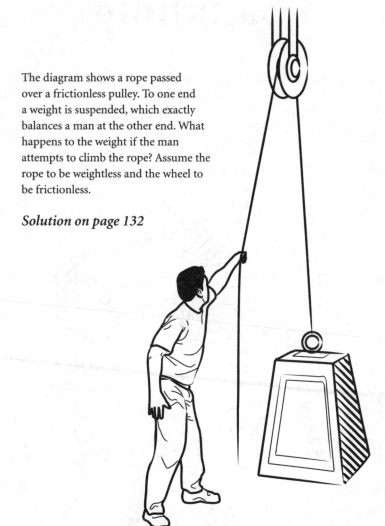

The tank

A closed glass tank is completely filled with water. A cork sphere is fixed to the bottom by a thread to prevent it from rising further. A steel sphere hangs from a thread fixed to the top of the tank. The lengths of the threads are such that the steel sphere is hanging just above the cork sphere.

What happens if the tank is suddenly moved to the right?

Solution on page 133

Can you stop an echo?

When something makes a loud bang between us and the side of a house or a cliff, we hear it and its echo. What would we hear if the source of sound were touching the surface?

Solution on page 133

The bottle and the coin

A well-known parlour trick is shown below:

A small coin is put on a card (e.g., a visiting or business card) and placed over the mouth of a bottle. When the card is flicked away with the finger, the coin drops into the bottle.

Can you explain why this should be so?

Solution on page 133

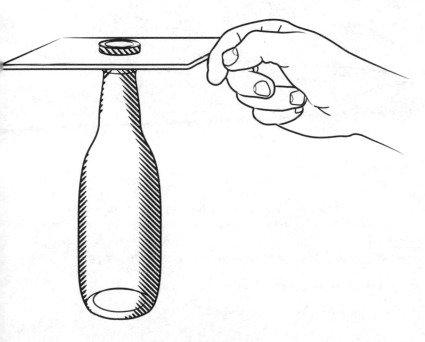

Sand on the beach

Walk along a beach at low tide when the sand is firm and wet. At each step the sand immediately around your foot dries out and turns white. Why? The popular answer, that your weight "squeezes the water out," is incorrect: sand does not behave like a sponge. So what does cause the whitening?

Solution on page 133

What's in the box?

The box illustrated is resting on a table, projecting slightly more than half its width over the edge. It can remain in this position owing to the weight of something inside it. When the table is bumped vertically, the box falls. What does the box contain? Its contents can be found in most households.

Imagine a similar box, but projecting less than halfway over the edge of the table. In this case it can be made to fall without bumping, touching, or applying any pressure to it or to the table, directly or indirectly. How? What is in this box?

Solution on page 133

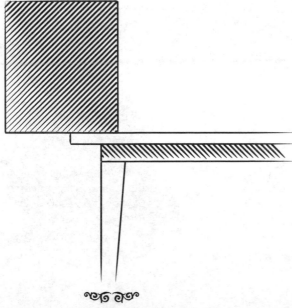

Quito

Suspend your disbelief now for this most intriguing puzzle. Approximately how much would a wman living in Quito, Ecuador, weigh if she were 70,800 kilometres (44,000 miles) tall? Would it be more or less than a thousand tons? (This needs a great deal of thought, not just an inspired guess.)

Solution on page 134

How do you get an egg into a bottle?

To put big objects into bottles has become something of an art. It seems as impossible as passing a camel through the eye of a needle, but these objects include such items as pears, arrows, and ships. Various techniques are used that might differ from object to object.

Our puzzle deals with a classic of this type, namely a hard-boiled egg into a milk bottle.

The method is pretty well known. First you drop a piece of paper and a lighted match into the bottle, then place a peeled egg upright over the neck. The oxygen consumed by the fire creates a partial vacuum and the atmospheric pressure will push the egg into the bottle. After you wash the ashes and dead match out of the bottle, you can take a bow.

But, can you do the same with an *unpeeled* egg?

Solution on page 134

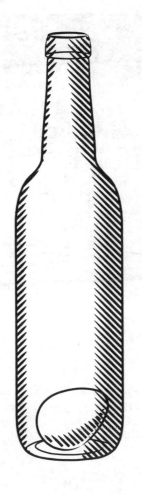

How does the Moon cause tides?

The motion of the Earth and Moon causes the tides, and the ceaseless sloshing of the tides is having an effect on the motion of the Earth. What is happening?

Solution on page 134

How do you photograph the Moon?

If you want to take a picture of the full Moon and get the largest possible image on film, should you shoot it when it is directly overhead (and therefore at its closest point to your position on Earth), or when it is down hear the horizon? Almost everyone says that the Moon is largest near the horizon. Is this an atmospheric effect or a psychological one? Does it show up in photographs?

Solution on page 134

How long to travel through space?

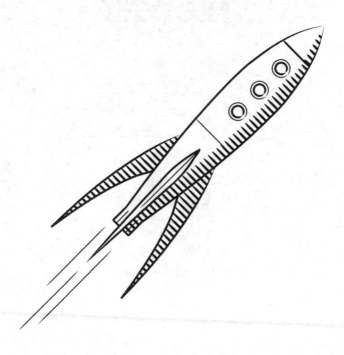

The nearest star to Earth is Alpha Centauri, which is about 4.3 light-years away. Assuming we can travel at the speed of light, what determines the amount of time our intrepid voyagers would need to reach the star?

Solution on page 134

What's in the middle of a rainbow?

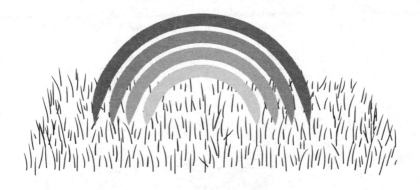

The rainbow is one of nature's most uplifting spectacles. Occurring as it does when sunshine meets suspended water droplets, as after a summer shower, the rainbow is a universal symbol of optimism. We most often think of a rainbow as an arc, round on top with legs below, only because water droplets are rarely seen below the horizon. You can see full-circle rainbows if you stand near a waterfall or a lawn sprinkler, or if your vantage point is "above the weather," as from a cliff or the top of a tall building. When you see a more familiar arc rainbow, imagine extending the arc's curvature into a complete circle. What feature, then, will you see at the circle's centre?

Solution on page 134

Climbing the mountain

We were all climbing the local mountain, and as we got higher it seemed to get colder and colder. John said it was due to our greater exposure to the cold winds. Bob said it was something to do with the more rarefied atmosphere. Tom said it was just imagination and as we were nearer the Sun it must really be warmer. Bill said we were farther from the centre of the Earth, which was known to be hot. What do you say?

Solution on page 135

The flawed sense

The five faculties through which we perceive the world—sight, smell, taste, hearing, touch—are, together with the brain, a most complex communication network, unsurpassed by any computer system created by man. However; one of our senses is flawed, and it is that flaw which enables us to enjoy a leisure activity which dominates many people's daily life. What is it?

Solution on page 135

How do you skim stones?

The ability to skip flat stones across the water is a question of skill. It is difficult to measure the path of a stone across water, but if you skip a stone on the wet sand at the water's edge, it will leave marks tracing its path. The flight is surprisingly complex. Long hops of several feet alternate with short hops of just a few inches, and zigs to the left alternate with zags to the right. A right-handed throw, with the proper grip, spins clockwise and strikes the sand first with its trailing edge. Will the first hop be short or long? To the left or the right?

Solution on page 135

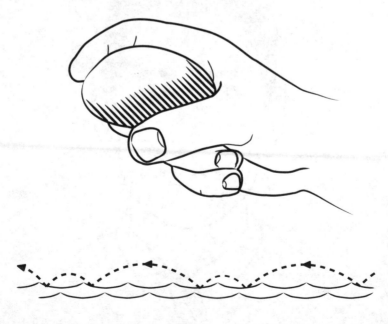

Rays of the Sun

When conditions are right, the Sun's rays streak across the sky from behind a distant cloud or mountain. Meteorologists call these crepuscular rays. They always fan out from a point that seems to be just behind the obstruction.

But wait, aren't the Sun's rays supposed to be parallel when they reach Earth? How does a cloud or mountain cause the rays to diverge that way?

Solution on page 135

The space station

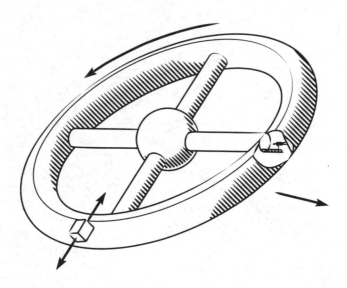

Weightlessness in a space vehicle is highly inconvenient to any astronaut in many ways. For example, he cannot pour liquid into a cup, neither can he drink from it: controlled movement is possible only by the use of handrails and so on.

It has been suggested that the space stations of the future for use of manned observatories or as staging posts for space exploration might be built in the form of huge wheels with hollow rims, as illustrated. These would be set in rotation so that the outer rim, which acts as the floor, would apply a radial centripetal force to the occupants or any objects inside to keep them moving in a circle.

The equal and opposite reaction to this centripetal force which any person or object exerts on the floor would act as an artificial weight, thus allowing eating, drinking, and working in comparative comfort. The quantum of this weight could be made equal to or less than the normal Earth weight simply by adjustment to the speed of rotation…

1. What would happen to an astronaut's weight if he were to walk round the space station in the direction of its rotation and then turn round and walk in the opposite direction?

2. Now suppose you are in a small windowless room in the same space station after suffering a bout of amnesia. In other words, you do not remember that you are in space. The speed of rotation around the hub produces a simulated gravity of one g. Inside your room everything seems "normal"—gravity seems to be operating on you exactly as it would on Earth. In fact, as far as your senses tell you, you are on Earth.

In your pocket you have a magnet, a piece of string, some coins, a pencil and a steel paper clip. Suddenly you have some doubt as to where you are. Is there a simple test you could do in your room, using one or more of the objects in your pocket, which would confirm that you are on a spinning space station and not on Earth?

Solution on page 135

An even playing field?

The World Cup playing field was painstakingly constructed to be a perfect plane. During the national anthem the 22 players, the referee, and the 4 linesmen were standing at attention. However, it was unlikely that any one of them, but at best only one, actually stood upright. Explain.

Solution on page 136

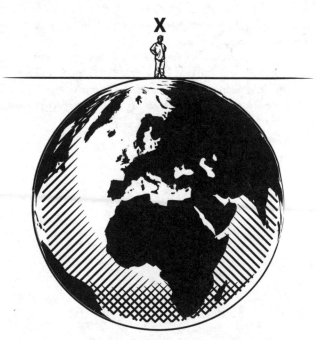

How do you tell acid from alkali?

Litmus paper is widely used for laboratory purposes. It consists of an absorbent paper impregnated with a dye, and is the oldest form of acid-base indicator. Dipped into an acid solution it will turn red, and in a base solution it will turn blue.

If no litmus paper is available, do you know a natural product which will serve the same purpose?

Solution on page 136

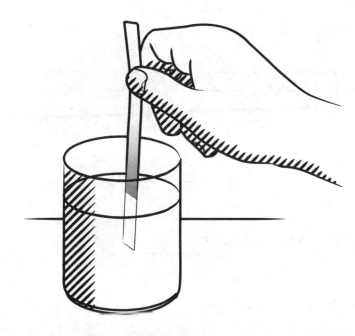

How best to build a bridge of bricks?

A farmer wants to build a bridge made of bricks, over a small river. His son, a student architect, suggests two designs as shown:

1. Which design can carry heavier traffic?
2. Which will need more building material?

Solution on page 136

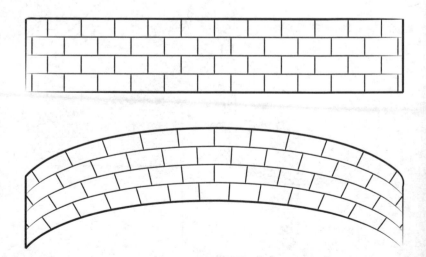

Whisky and soda?

You have a quart of soda water and a quart of whisky. Take one tablespoonful of soda, transfer it to the whisky, and mix thoroughly. Then take a tablespoonful of this mixture and pour it back into the soda. Is the amount of soda in the whisky greater than, less than, or equal to the amount of whisky in the soda?

Solution on page 136

Can you see 3D?

Dot stereograms specially designed patterns, usually printed in colour, that offer an interesting illusion of 3D depth. Once the viewer has mastered the technique, the effect is spectacular.

It is, of course, one thing to enjoy the images, but quite another to understand how it works. Can you find an explanation for the 3D effect of the patterns, and prove that your theory is correct?

Solution on page 137

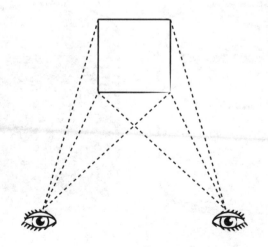

The wheelbarrow

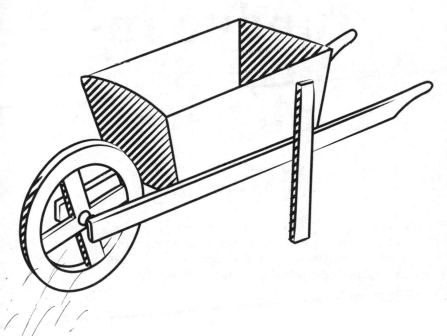

Which is easier: pushing a wheelbarrow or pulling it?

Solution on page 137

Tongue-in-cheek physics

This section includes ideas that might be theoretically feasible but, due to real-world limitations, may not be realized. Other items are fallacies, or contain flaws in reasoning that are only too obvious. As such, these problems are more amusing than intellectually challenging.

Air Mail?

When posting a parcel, you have to pay more if the parcel is heavier. There is a scale of charges linked to the weight of the parcel. The post office is quite right to charge according to the mass of the parcel. However, I have noticed that the machines used are spring balances, which measure weight rather than mass.

This means that I could employ the following scheme. When I wanted to send an object through the post, I could enclose it in an oversized box, which would leave space for me to put a helium balloon in with it. The parcel has the extra mass of the helium and the fabric of the balloon, but because of the upthrust on the balloon the parcel would weigh less, saving me postage.

Presumably, if the balloon were large enough, the parcel would have negative weight, and the post office would have to pay me to transport it. Would the scheme work?

What machine would correctly measure the mass of the parcel?

Solution on page 138

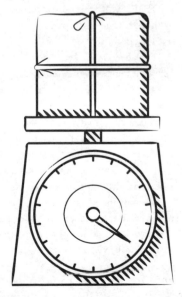

The Spying Game

In war as in industrial espionage, the main problem is to avoid detection when transmitting messages. Microdots and invisible ink are old hat, and new methods are devised all the time.

Paul Goodwin was employed on a top-secret project concerning isotopes and his problem was to transmit a message to his principals abroad. Coded messages were too dangerous and therefore unthinkable. After much consideration he devised what he considered to be a foolproof method.

There are 26 letters in the alphabet. As any message would also contain punctuation marks and blanks (between words), we are dealing with a total of about 40 symbols. Paul assigned a two-digit number to each, starting with 01 for "A", 02 for "B", and so on until he reached 38 for a full stop and 39 for an interval between words. (ISOTOPE MASS would, for instance, read: 09 19 15 20 15 16 05 39 13 01 19 19.) With the help of his powerful computer, Paul Goodwin translated the message into one long number.

The reader is now asked to permit some intellectual licence. By putting a decimal point before the number, we can consider the whole code to be expressed as the length in metres of a bar made of precious metal (to avoid any chemical reaction with the atmosphere). This bar would have a length of more than 9 but less than 10 cm.

Let us assume that Paul has the means to cut the bar to the required length, which would require fantastic precision.

Let us also assume that the recipient of the bar, having equally precise measuring equipment, would obtain the same decimal number and decode the secret message.

Assuming that the precision needed for this operation were available, would it work, at least in theory?

If not, can you think of a procedure which would make it theoretically viable?

Solution on page 138

How heavy can you get?

Let us suppose that there is no limit as to how tall a person could grow. Now imagine someone growing rapidly, becoming progressively taller and therefore heavier. The question the reader is asked to consider is this: Would the person's weight increase forever?

Solution on page 139

Can you float a battleship in a bathtub?

Specifically, the battleship below – which, floating on the high seas, has a mass of 30,000 tons. Imagine slowly lowering the ship into a huge bathtub that is shaped like the ship but a little larger, containing a small volume of water. As the ship is lowered, the water is forced up the sides of the tub and over the rim until there is only a thin envelope of water between the tub and the ship's hull. Is it possible to float a 30,000-ton battleship in just a few hundred gallons of water? Keep in mind that a floating object displaces a volume of water equal to its weight and that 30,000 tons of water have a volume of several million gallons.

Solution on page 139

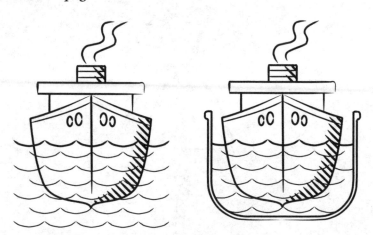

The light and the shadow

A man is walking along a level road lit only by a solitary streetlight. He is moving at a constant speed and in a straight line as he passes the lamppost and leaves it behind, which makes his shadow lengthen. Does the top of the man's shadow move faster, slower, or at the same rate as the man? Prove your answer.

Solution on page 140

Feathers and Gold I

Which weighs more, a pound of feathers or a pound of gold?

Solution on page 140

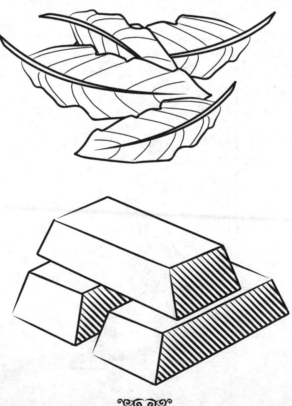

Solutions
& Explanations

How do I blow up my stereo?

Even with correctly connected speakers you find that in most parts of the room there will be several frequencies of sound that combine destructively while others combine constructively. This means that some frequencies will be quieter than they should be, while others are louder; but this is not normally noticeable among the thousands of other frequencies present. At any point equidistant from each speaker, however, you can guarantee that all the frequencies will combine constructively, giving the best listening position.

But if you wire one of the speakers with the reverse polarity, at the central position all frequencies combine destructively, giving the worst possible listening situation.

How can I turn red to green?

If you filmed the LED with a high-speed camera to slow down the effect, you would see it flash red and green. But our eyes, by a process called 'persistence of vision', work much more slowly and you would be unable to see different-coloured flashes.

Also, our eyes do not have a colour receptor for each colour in the rainbow. There are only three types of colour receptor in our eyes, sensitive to mainly red, mainly green, and mainly blue. The other colours are detected by triggering more than one receptor to a greater or lesser extent. Colour televisions make use of this fact; if you look closely at the screen you will see that the picture comprises only red, green and blue dots. Our flashing LED will stimulate the red and the green sensor simultaneously, giving the sensation of yellow light.

Why does the ringing change?

There was a strip of railing nearby. Sound reflected back from each individual support rail arrived as a series of small echoes separated by a small interval of time. This caused a distinct ringing tone, the frequency of which was determined by the distance between the rails.

Why can't I tune my radio?

There would be even worse problems if two or more transmitters attempted to broadcast the same program on the same frequency.

As radio is a type of wave, at some points the two waves from the two transmitters could just happen to be synchronized so the radio would pick up a very strong signal. At other places the waves could be completely out of synch and would cancel each other out. Thus, as you moved around, the radio would continually go from extremely loud to no signal at all within the distance of a few metres.

Are there rainbows on the Moon?

Part of the mysterious beauty of the rainbow:

1. You can see the spectrum only if the angle of refraction between the Sun, the drop of water, and your line of vision is between 40° and 42°.
2. Lunar rainbows are also possible but rare, because moonlight is not as strong as sunlight and its intensity varies with the phases of the Moon.
3. This phenomenon is called dew-bow and is caused by the water drops on the grass.

How do I dull my headlights?

It would work. However, it was not introduced because the disadvantages would outweigh the benefits.

1. Not being able to see the headlights of approaching cars would be dangerous, particularly when visibility is reduced during heavy rain or fog.
2. The windshield's polarization would absorb a fair amount of light in the street scene, reducing visibility generally.
3. At any rate, it would be difficult to provide an effective filter for windshields in view of their irregular shapes.
4. Coloured stress patterns would become visible in the wind-shield and this would be distracting for the driver.

How do I make a toy boat?

As the steam enters the cooler tubes it condenses, creating a vacuum that will draw in water. However, such water as is sucked in will not do so in the form of a jet, but will come from all directions and will not therefore have the same propulsive force.

Does wind affect my mind?

There are three questions to be answered:

1. Why is the wind warm? The meteorological conditions have to be right for the cold

mountain air to sink in the form of a fast-moving wind. As the air sinks it encounters the higher pressures found at lower altitudes and is compressed. This compression causes warming; the result of a phenomenon known as adiabatic process (see Glossary).

2. Why is the wind dry? The air was originally cold and cold air can hold less moisture than hot air.

3. Why does it have psychological effects? Nobody is sure that it does, or if so what the reason might be.

Why is the safety lamp safe?

The explosive gases would still infiltrate through the mesh and ignite inside the lamp. The screen would quickly conduct away the heat, thereby preventing the flame from escaping to cause an explosion. The change in the flame also gives a warning of danger to the miner.

Can a mirror turn me upside down?

The answers are:

1. Place the mirror on the floor (or ceiling).
2. Place the mirror at a 45° angle to the floor.
3. I cannot think of one either.

Can you square a circle?

The apparent gravity increases the farther one moves from the point of rotation. The gravitational effect would therefore be greatest at the edges of the room. For example, a marble placed in the centre of the floor would roll towards one of the two edges, which are parallel to the axis of rotation.

Even though the floor is perfectly flat, it would feel as though there were a hump in the room, that one would be walking 'uphill' towards its centre.

Why does the Sun go out?

There are two reasons why solar eclipses are so rare:

1. The Moon's orbit is not in the same plane as the Earth's orbit around the Sun. This means that every month when the Moon is on the correct side it is usually too far to the north or to the south to block the Sun's light.

2. If the two orbits were in the same plane, a total eclipse would occur every month, but it would still be seen from very few places on Earth. This is because the Moon's shadow, when cast on the Earth, is only a few miles across. Only if observers happen to be in this shadow area will they experience a total eclipse.

Why aren't I floating?

There is no way to distinguish between 1) gravity and 2) acceleration. However, in a rotating spacecraft, 'gravity' will appear greater farther away from the centre of rotation, so there are several things that he could do:

1. Weighing an object with a spring balance will give a smaller reading near the ceiling.
2. A marble placed in the centre of the floor will roll towards the edge.
3. Two plumb lines will diverge rather than hang parallel.
4. He would not be able to spin a coin.

Pull of the Sun

Let us consider a simpler system first. Suppose there were only the Earth and the Sun. If the Earth were stationary in space, the two would be drawn together by their mutual gravitational attraction. However, there is a more stable possibility: the Sun's gravitational force is used to maintain the Earth's orbit rather than drawing it closer.

The same process explains why the Moon orbits the Earth. The Sun's gravity is used to maintain the Earth/Moon system in orbit. If all orbiting suddenly ceased, all the planets and their moons would instead fall towards the Sun.

Why do I fly back to Earth?

At first the atmospheric drag is very small and has the effect of making the satellite fall into a lower orbit. The speed in orbit is determined by height: the lower the orbit the faster the speed. The satellite therefore travels faster.

This may seem like a puzzling contradiction to the law of conservation of energy. But the total energy of the satellite, potential and kinetic, has in fact decreased by the amount of heat produced by friction.

Eventually, as the orbit spirals downward, the frictional drag will be great enough to slow down the satellite as it falls into the main body of the atmosphere. The increase in orbital speed is similar to the effect on a spinning ice-skater who pulls his outstretched arms nearer to his body, thereby increasing his speed of rotation.

What colour is the Sun?

We can think of five answers:

1. White – almost by definition, the colour of sunlight is white.
2. All colours – one could argue that there is no such 'colour' as white. Isaac Newton discovered that white light is really a mixture of all visible colours.
3. Yellow – the most abundant colour in sunlight.
4. Black – the Sun appears white because it is emitting light, rather in the way a light bulb looks white while it is on; the bulb element is gray while the bulb is off. But what

colour would the Sun be if it cooled down? Scientists know that the Sun is close to being what is called a Black Body Radiator. So if it cooled down it would be black.
5. No colour at all – black is not really a colour but is the absence of all colours.

Why isn't the whole sky as bright as the sunlight?

Olbers himself provided the counter-argument.
1. While light from some of the galaxies we observe left them millions of years ago, the light of many millions of stars has not yet reached us. Our own galaxy is more than one hundred thousand light-years across.
2. The paradox assumes that all or most stars in the observable universe are alight at the same time. This is a misconception, because the life span of a star is limited to about 10^{10} years. Although this is a long time, it is still finite.
3. The universe itself is probably not infinite.
4. Light may also be absorbed by dust, and dark matter (if it exists).

Why don't I burn my head?

I had dried my ears with a towel and therefore they felt the real temperature of the air. My hair was still wet, so the fast-moving air was rapidly evaporating the water. Evaporation produces cooling (this is the principle used in the domestic refrigerator), and therefore the air felt cool on the wet hair.

So near and yet so far?

If a telescope (or a pair of binoculars) is being used in the correct way, it is said to be in 'normal adjustment'. When utilized in this way, the eye is perfectly relaxed when using it. When a perfect eye is relaxed, it focuses at infinity. A telescope does not, therefore, give you an image that is closer. It does, however, make the object look bigger.

A magnifying glass increases the focusing power of the eye, allowing the object to be placed closer. There are many circumstances in which it works by not magnifying at all. Try this simple experiment to see what I mean. Hold a piece of writing so close to the eye that it cannot be seen clearly because it is blurred. Now insert a magnifying glass between the eye and the writing. You should now find that the writing is clear enough to read but is no larger than it was.

Glaringly obvious?

Strangely enough, the answer to this question is yes. If there are just two crossed polarizing filters, no light will pass through the combination. If you insert an extra filter between the other two, at 45 degrees to them, then some light will travel through this triple combination. In fact, the amount of light transmitted through the triple combination is about half of what would pass through a single filter.

This is because some light will always manage to get through a double combination at 45 degrees to each other. So some light will pass through the first and second filter. Some light will also pass through the second and third filter. Therefore, some light will pass through the complete triple combination.

How can I balance my scales?

You might think that the correct result is the average, 122 grams, but this would be true only for a balance with equal arms. This balance must have unequal arms, and the formula for this is the square root of the product: 100 x 144 = 120 grams

Proof: If the scale's balance when empty is inaccurate, the two arms must be of different lengths (a, b).

Let the corresponding readings for the object X be W1 and W2

At balance:
$aX = bW1 \rightarrow a/b = W1/X$ and $aW2 = bX \rightarrow a/b = X/W2$

Therefore:
$W1/X = X/W2$ or
$X = SqRt (W1 \times W2)$

Substituting:
$X = SqRt (144 \times 100) = 120$ grams

How can I tell if an egg is raw?

If you spin the egg it will stand on end if it is hard-boiled, similar to a top. A raw egg is unstable because the contents are viscous, and therefore it will not spin.

Why does a golf ball have dimples?

In fact, dimpled golf balls travel about four times farther.

The effect is caused by the backspin imparted to the golf ball by the club. As the top of the ball spins backwards, it drags air which would otherwise have traveled beneath the ball. This air has to speed up to travel the extra distance, and this causes lift in a very similar way to an aircraft wing. If the ball were smooth, the spin would have no effect.

How do you cool off?

It is a mistake to believe that by simply opening the refrigerator door you will achieve your objective. You might momentarily feel cooler, but you will soon increase the temperature as the motor releases more heat than it reduces in the refrigerator.

The only way to succeed is to switch the motor off and open the refrigerator door. The cold air will mix with the warm air and achieve an overall reduction.

How can I make a spray gun?

The air travelling through the blow tube at considerable speed reduces the air pressure in the conduit tube. The liquid in the container is subject to atmospheric pressure and is therefore forced up, and out of, the conduit tube in the form of a fine spray.

What makes glue stick?

Molecules of a substance attract each other by a process called cohesion. Molecules will also attract molecules of a different substance, and this process is called adhesion. The trick is to find substances whose molecules are very adhesive.

To work effectively the molecules have to be in close proximity, which is why adhesives are usually liquid and the surfaces involved should be as clean as possible.

Rather surprisingly, water is quite a good adhesive. If, for example, you wetted two pieces of wood, placed them together and put them in a freezer, once frozen the pieces of wood would be very difficult to part.

Theoretically, if one could hone two surfaces to a degree that molecular contact was possible, no adhesive would be needed. In practice such surfaces are contaminated by dust and other impurities, so that adhesives are needed.

What's that strip?

The metal strip is designed to ground the vehicle. Rubber is a good insulator, and it is possible for a car to build up a charge of static electricity.

There are several reasons why people might want to use such a device. Some worry that this static charge could cause a spark which could ignite gasoline vapour, causing a fire. It is possible that the strip would avoid such a spark, but it is by no means proven that there is a fire risk in the first place.

Other people use it to avoid an electric shock as they close the car door on leaving. This static build-up is caused by the rubbing of their clothes against the fabric of the seats. The metal strip is unlikely to solve this problem; not wearing rubber-soled shoes would probably be more effective.

Some people have a theory that travel sickness is caused by static electricity. Any improvement in the situation here is likely to be more psychological than real.

How do you drive on ice?

Good drivers know:

1. To increase friction where it matters you will use the blanket in front for a front-wheel-drive vehicle, otherwise in the rear.
2. The lower the torque the better; therefore, you should start in second gear at slow speed.
3. To stop skidding you should turn your front wheels into the skid.

How do you float a balloon?

Attach an overlong piece of string to the balloon and hold the free end. The string will then form a curve, travelling down from the hand to a low point and then rising up to the balloon. Cut the string at its lowest point. The balloon will then be left with exactly enough string to cause it to hover.

When the leaves fall

Thousands of years of evolution have ensured that trees shed their leaves at the optimum time. If they drop their leaves too early, then valuable time would be lost. If the leaves fall too late, they could be damaged by frost, and this would be detrimental to the plant as a whole.

Why don't the clouds move?

As the wind flowed over the mountain, it rose and then descended. This wave continued for a few oscillations over the plain.

The conditions were such that, at the top of the wave, the air cooled enough for the water vapour to condense, forming a cloud at that point. As the air continued along its path, it sank and warmed sufficiently for the water droplets to evaporate again. This caused a series of clouds which were being continuously created at the back and destroyed at the front. The clouds were therefore stationary although the air was moving.

How do I bisect an object?

For symmetrical objects the bisectors will be coincident, but this is not true for asymmetrical objects.

Why can sheep still eat grass?

The evolutionary process can be much more subtle than one realizes. Unlike most other plants, grass grows from the base of its stalk and not its tip; it can therefore cope with grazing fairly successfully. If all the grazing animals were eliminated, other plants would then be much more successful while grass, being a small plant, would lose out to those other plants in the competition for light.

You just have to think of environments where there are large numbers of grazing animals to realize how successful this strategy is for grass.

How do you walk on a tightrope?

Walking a tightrope is possible only if you keep your centre of gravity precisely above the rope. This is a bit like keeping upright on a stationary bicycle.

The balancing pole is deliberately heavy, with weights placed in the tips. It makes things easier in two different ways. The bar has a large inertia, and by moving it sideways the walker can adjust his position. Also because the bar is heavy at the ends and bends downward, this has the effect of lowering the centre of gravity of the walker. If this could be lowered to below the rope, then the walker, although still looking very precarious, would in fact be very stable.

Should I run through the rain?

Like lots of real life problems, this is more complicated than it first appears. So at first, let's make some simplifying assumptions:

1. There was no wind – the rain was falling vertically.
2. I was wearing a hat – so we are considering the rain that I walked into, not the amount falling on my head.
3. The rain was falling at a steady rate.

Now let's think about the space that I moved through; we could imagine an invisible me-shaped tunnel stretching from where I am standing to my front door. Rain is continually entering this tunnel from the top, and leaving at the bottom. As it is raining steadily, there is as much rain entering the tunnel as leaving, so an equivalent system would have the raindrops hovering stationary. It is now obvious that however quickly I move though my personal tunnel, I will collect exactly the same number of raindrops.

But remember, the rain stopped as I reached my doorstep, so if I had not run, I would have got through only about half my tunnel before the rain disappeared from the other half. By running I had actually become wetter than if I had continued my leisurely walk.

If the rain had not stopped but had continued falling steadily, I would have been just as wet either way. However, there is an element of chance; if it had started raining more heavily, I would have got wetter by walking.

Now suppose I had not been wearing a hat. The top of my head continues to get wet as long as I am in the rain. Running keeps my head drier.

If there had been a head wind, this would have made me wetter the longer I stayed out, i.e., continuing to walk would have made me wetter. If there had been a tail wind, running would have decreased the wetness to a point where the speed of the running equalled the wind speed, when my body would not have got wet at all!

So, is it better to run? It depends....

How do I stop my tyres wearing out?

My friend was correct in assuming that if a car travels in a circle the outside tyres travel farther and therefore wear more than those on the inside.

However, it is a surprising fact that the extra distance traveled by the outside wheel is not affected by the radius of the circle around which the car is travelling. So his new way of driving to and from work would succeed in equalizing the wear on the tyres.

Why don't planes fly lower?

Even taking the extra distance into account, it is still cheaper to fly at a higher altitude where the aircraft is flying above the turbulent weather patterns. Also, the thinner air presents less drag, meaning less fuel is consumed. Finally, it may be possible to take advantage of the fast winds found at that higher altitude.

How can I sail the world?

If they both set off at the same instant and also returned at the same time, then both their journeys would have taken the same time. However, the two ships would have counted different numbers of passing days. If the island had experienced the passage of X days, the ship travelling west against the rotation of the Earth would have counted X-1 days because it would have made one less rotation than the world. The ship travelling with the spin of the globe would have counted X+1 days.

This anomaly was corrected by the adoption of the International Date Line.

Does the Earth move?

There are two ways in which earthquake waves can travel around the world. First, they can travel across the Earth's surface, rather like waves travel across the sea. Second, they can travel directly through the centre of the Earth. These waves travel at different speeds and have different distances to travel, so they arrive at the seismograph at different times.

The Earth does not have a uniform density. The denser core acts like a lens, focusing the vibrations at some points while leaving other places in a shadow.

How do you burst a barrel?

The trick works because the water pressure at one point depends on depth, not the weight of water above. If the tube connected to the top of the barrel is narrow, a small amount of water can cause a large increase of pressure in the barrel by increasing its depth below the surface.

If a narrower tube were used, then even less water would be needed to increase the depth. If the tube were half the area of cross section, then half the amount of water would be needed to attain the same height in the tube. So half the jug would be sufficient.

Why doesn't the snow fall evenly?

The wind driving the snow diverges many metres in front of a large building, thus dispersing the snow before it hits the wind-side face. A smaller object does not divert the wind, permitting the snow to build up.

How do I shoot in a straight line?

You should aim to the left of the target. Any moving object will be deflected to the right north of the Equator, and to the left south of the Equator, in relation to the rotation of the Earth.

This phenomenon is called the Coriolis effect, named after the French physicist Gaspard de Coriolis (1792-1843), who first analyzed it mathematically. The Coriolis effect is of great importance to meteorologists, navigators, and the military (*see* **Glossary**).

How do you drink on ice?

Polar ice does indeed contain salt. Such ice, melted down, is as undrinkable as seawater. However, over time, the brine in the ice blocks will migrate downward, because of gravity. This draining effect will make the melt-water drinkable after about a year, and it will be almost completely free of salt after several years.

The problem does not affect all polar regions, as some ice is formed by precipitation. However, many areas have insignificant snowfall, and the little there is gets blown away by winds of up to 100 mph.

Among other desalination techniques, freezing salt water has been developed as an alternative method, based on the different freezing points of fresh- and seawater, but the equipment needed is beyond the reach of Eskimo communities.

Why do boomerangs come back?

The return boomerang has a length of 30 to 75 centimeters (12 to 30 in), curving to the left, and capable of more than 90-metre (300 ft) throws. There have been several attempts at explanations. T.L. Mitchell in 1846 suggested that it was caused by the skew combined with the spinning motion. This does no more than beg the question. A more convincing explanation is offered by E. Hess in *The Aerodynamics of Boomerangs* (*Scientific American*, Nov 1968). According to Hess, boomerang is an aerofoil and therefore subject to lift, which is greater on the top half because it is turning in the same direction as the boomerang, whereas the bottom half is turning in the opposite direction.

How did we avoid rockets?

The V-2 rockets exceeded the speed of sound and therefore you could not hear their approach. You could hear only the detonation after they hit the target, too late to take evasive measures.

How do you steer a boat?

The rudder will have an effect only if there is relative motion between the boat and the water. This answer begs the question of whether there is such relative motion? Most likely yes. However, there are so many forces acting upon the boat that this question cannot be answered with absolute certainty.

First, there is gravity. Imagine the river to be frozen. Ignoring friction, the boat would slide down the slope. Then there is buoyancy, resulting in a component force downriver, partly counteracted by drag. Air resistance adds to the drag, while a wind upstream or downstream has an effect. Furthermore, the river flows at different rates in the middle and near the banks.

It is theoretically possible that all these forces combine to synchronize the motions of the boat and river for a limited period, which would make steering impossible.

Can you fly faster than sound?

1. There is a common misconception that a sonic boom occurs only at the moment when a plane exceeds the speed of sound. In fact, any aircraft travelling faster than sound creates a large pressure disturbance that travels along with the plane. If this pressure disturbance passes a person on the ground it is detected as an explosive sound. That is why the Concorde is not allowed to fly supersonically over land. As this pressure wave does not pass the passengers inside the plane, they are unaware of the sonic boom.

2. No. Breaking the sound barrier is a common expression meaning travelling faster than sound. Sound does not always travel at the same speed; it is affected by the air

temperature. As air is colder at high altitudes, the speed of sound is correspondingly lower by about half a metre (20 inches) per degree of Celsius. At 0°C, the speed of sound is about 328 metres (1,075 ft) per second, and at 16°C about 338 metres (1,108 ft) per second.

Why is Death Valley so hot?

A number of factors contribute to these extreme conditions. The mountain range on the west side rising to more than 3,300 metres absorbs the moisture of the west winds, which, on descending east of the Rockies, are adiabatically heated and dried, turning the valley into a hot desert.

The Panama Canal conundrum

The two responses desired are:

1. The canal is fed by a number of freshwater lakes including Gatlin and Miraflores. When the last gate leading to the ocean is opened, the freshwater level will still be higher than the denser saltwater, creating a flow to equalize levels and providing a drift for the vessel.
2. The salinity of the Pacific is higher than that of the Atlantic, and therefore denser, which accounts for the lower level of the Pacific.

Balloons in the bath

It will remain where it is placed.

Why don't submarines rest?

Water pressure pushes perpendicular to a submarine's hull at every point, at the bottom as well as at the top and sides. When a sub settles on a clay or sandy bottom, the water layer may be squeezed out from beneath the hull, robbing the sub of much of its upward buoyant force. In effect, the downward forces can then glue the sub to the bottom.

Does a bootstrap elevator work?

Although it looks as if the man were trying to lift himself up by his own bootstraps, he really isn't. True, for every pound of force that he pulls up on the rope, he also pushes down an equal force on the block, but if he is strong enough to lift his own weight plus the weight of the block, he will rise from the ground. (Tests have shown that a 190-pound man can lift both himself and a 110-pound block in this way.)

Does the balloon move?

As the car accelerates forward, the balloon on its string tilts forward, too. Inertial forces push backwards in the car, pressing the people against their seats (an effect with which we are all familiar), but also compressing the air at the back of the closed car. This increased air pressure at the rear pushes the balloon forward. For similar reasons, as the car rounds a curve, the balloon tilts into the curve.

Can you make a boat on land?

It all depends on how the oval wheels are affixed to the vehicle's axles. If the wheels on the opposite ends of the same axle are positioned at right angles to each other, a roll will be produced. By synchronizing the front and rear wheels, so that on each side of the vehicle the two wheels have their long axes at 45-degree angles, the carriage will both pitch and roll. If, on each side of the vehicle, the two wheels also have their long axes at right angles, the carriage will merely move up and down alternately on two diagonally opposite wheels.

These possibilities leave only the problem of finding a driver prepared to put up with any of them.

What makes a mirror reflect?

Left and right are directional concepts while top and bottom, or up and down, are positional concepts. The same incidentally applies to east-west and north-south. Walk northward along the Greenwich Meridian and Berlin will be to the east and on your right (ignoring the fact that you could also travel to Berlin the long way round) until you reach the North Pole. Crossing it, you turn around to still look north and Berlin will be to the west and on your left. Yet north and south will remain in the direction of the poles. Equally, up and down assume the centre of the Earth as reference point. An ordinary mirror will reverse direction, but not position.

The tall mast

They open the stopcocks and let some water into the boat. This makes it ride lower in the water so that the mast clears the bridge.

The spectrum

Pigments in common use are impure. Because of these impurities, yellow paint presents the eye with a mixture of red, yellow, and green, and blue paint offers a mixture of blue and green. In each case, however, the basic colour dominates so that the eye perceives only yellow and only blue respectively. However, when the two paints are blended, yellow paint absorbs blue, and the blue paint absorbs red and yellow light, leaving green as the only remaining colour common to both paints.

The colours of projected light, however, are pure, and if they are complementary colours, such as blue and yellow, or green and red, white light is the result.

In other words, when paints are mixed, the resulting colour is produced by absorption. When lights are mixed, the results are produced by combination.

How do you open a glass jar?

You will likely need someone's help with this: another pair of hands to hold the bottle firmly while you wrap a narrow strip of cloth around the bottle's neck. Then, by pulling the cloth rapidly to and fro, you can cause friction between the strip and the glass that generates sufficient heat to expand only the bottle's neck so that the stopper can be withdrawn easily.

Heating otherwise does not do the trick, as bottle and stopper would expand uniformly.

How do you make a truck fly?

If the bird is in a completely airtight box, the weight of the box and the bird will be the same whether the bird is flying or perching. If the bird is flying, its weight is borne by the air pressure on its wings; but this pressure is then transmitted by the air to the floor of the box. If the bird is flying in an open cage, part of the increase in pressure on the air is transmitted to the floor of the cage, but part is transmitted to the atmosphere outside the cage. Hence the cage with the bird will be lighter if the bird is flying.

Hole through the Earth

1. The ball's velocity would steadily increase from zero at point A to a maximum at the centre of the Earth. It would steadily decrease thereafter to zero at point B, taking 42 minutes for the complete trip. This fascinating speculation goes back to Plutarch. Even Francis Bacon and Voltaire argued about it. Galileo gave the authoritative answer which is generally accepted: The ball would fall faster and faster, though with decreasing rate of acceleration, until it reached maximum velocity, about 5 miles per second, at the Earth's centre. It would then decelerate until its speed reached zero at the far end of the hole. If air resistance is ignored, it would oscillate back and forth, like a pendulum, *ad infinitum*.

2. At the centre of the Earth, there is no overall gravitational force on the ball. This is because the Earth's pull on the ball will be equal in all directions. Therefore, the weight will be zero.

3. The weight will change but the mass will not.
4. The ball would be in free fall throughout the entire trip, so it would always be in a state of zero gravity.
5. More time. The trip would take about 53 minutes. Although the distance is much shorter than on Earth, the Moon's gravity is only about one-sixth of that of the Earth.

Feathers and Gold II

An ounce of gold weighs more. The Troy system has 12 ounces to a Troy pound, whereas in the Avoirdupois system a pound consists of 16 ounces. So a Troy ounce is greater than an Avoirdupois ounce.

The two Feathers and Gold puzzles demonstrate the vital importance of defining precisely the units in which a measurement is made. Units of measurement are objective, ultimately reflecting some measure in the world as observed by man. Thus, the measure of length is a foot, the measure of the size of a horse is the hand. Both the Troy system and the Avoirdupois system are based on the weight of a grain of wheat as developed by man, which (to use another system, metrics) is 0.0648 grams. A Troy pound is 5,760 grains, and an Avoirdupois pound is 7,000 grains. In the words of the first Greek sophist, Protagoras (fifth century B.C.): *Man is the measure of all things.* Whereas Protagoras meant this subjectively, to suggest that judgments are relative, it is better interpreted objectively, to express the nature of units of measurement.

The half-hidden balance

From 454 grammes to infinity. It is 454 grammes, not 1kg, because when the length approaches infinity the second weight converges to zero. When the hidden rod's length is zero the weight must be infinite to balance the force on the left.

How do you make salad dressing?

Since oil floats on vinegar, to pour mostly oil Jill had only to uncork and tip the bottle, then cork it, turn it upside down and loosen the cork just sufficiently to allow the desired amount of vinegar to dribble out.

Hourglass puzzle

When the sand is in the top compartment of the hourglass, the high centre of gravity tips the hourglass to one side. It is kept at the bottom of the cylinder by the resulting friction against the side of the cylinder. After sufficient sand has flowed into the lower compartment to make the hourglass float upright, the loss of friction enables it to rise to the top of the cylinder.

It is interesting to note that, if the hourglass is a little heavier than the water it displaces, the apparatus works in reverse. That is, the hourglass normally rests at the bottom of the cylinder, and when the cylinder is turned upside down, it stays at the top, sinking only after sufficient sand has flowed to eliminate the friction.

The toy is said to have been invented by a Czechoslovakian glassblower, who made them in a shop just outside Paris. Strangely enough, physicists seem to find this particular toy somewhat more puzzling than other people. They often advance sophisticated explanations which involve the force of the falling sand keeping the hourglass at the bottom. However, it is easy to show that the net weight of the hourglass remains the same whether or not the sand is pouring.

Can you blow up a room?

Nothing would happen because there is no oxygen in the room.

Rotation counter

In riding, the front (steering) wheel follows a more wobbly, therefore longer, course than the rear one.

Can you propel a boat with a rope?

By jerking on a rope attached to its stern, a small boat can indeed be moved forward in still water, and speeds of several kilometres per hour can be achieved. As the person's body moves toward the bow, the friction between the boat and the water prevents any significant backwards movement of the boat, but the inertial force of the jerk is strong enough to overcome the resistance of the water and transmit a forward impulse to the boat. The space capsule, in the absence of friction, cannot be propelled in the same way.

How many times can you tear a card?

Higher than the Sun is from the Earth, so more than 93 million miles.

The Light Fantastic

In one of the rooms, the light is provided by sodium lights, which provide a very pure yellow light. In the other room the light is provided by a mixture of red and green lights. Red and green lights, when mixed, appear yellow. In the first room, both red and green paint appear black under the pure sodium light, while in the second room the red and green paints are able to reflect their own colours.

Why do white horses survive?

The herds were plagued by vicious horseflies, causing severe debilitation of the animals. For some reason which modern science remains ignorant of, the horseflies showed a strong preference for the darker horses and did not attack the white ones. This resulted in the greater survival of white horses, so eventually giving rise to the famous white horses of the Camargue.

How does a helicopter fly?

Every revolving body develops a force acting perpendicular to the axis of revolution. This moment of force is referred to as torque, and would cause the helicopter to rotate out of control. The tail rotor is designed to counteract the torque.

In large helicopters with two rotors, the torque is avoided by having the rotors rotate in opposite directions.

The beaker

The answer is 2 cubic centimetres of water must be added to the left-hand beaker: one to balance the cubic centimetre of water added to the right-hand beaker and one to balance the buoyant force exerted on the iron bar due to its displacing one cubic centimetre of water when the water level has been raised one centimetre.

Drops and bubbles

The bubbles would attract each other. If water is removed from one spot in all space (bubble A), the gravitational balance surrounding it is upset, and the net effect on a nearby molecule of water is that it is drawn toward greater mass; that is outward, away from the bubble. If there are two bubbles, the water between them acts as if it is repelled from both, and the bubbles would move toward each other.

Why can't you copy a banknote?

No actual picture is made of it except on the television screen, where it is emergent. The recording on tape is a linear message or code, not a recognizable picture.

Rope and pulley

The answer is that, regardless of how the man climbs – fast, slow, or by jumps – the man and weight always remain opposite. The man cannot get above or below the weight, even by letting go of the rope, dropping, and grabbing the rope again.

The tank

If the tank is suddenly moved to the right, inertia will cause the steel sphere to move to the left relative to the tank. The steel sphere tends to persist in its state of rest. The water tends to do the same, but the sphere, being heavier than water, dominates. Because the cork sphere is lighter than water, it moves to the right relative to the tank. If the tank were moved back and forth, the two spheres would also move back and forth: the steel sphere in the opposite direction, the cork sphere in the same direction.

Can you stop an echo?

You'd hear no recognizable echo, which is a reflection of sound, but the decibels (intensity) would be doubled, much as is candlelight when placed next to a mirror.

The Bottle and the Coin

Isaac Newton's first law of motion explains this experiment:

Every body continues in its state of rest or of uniform motion in a straight line, unless compelled by some external force to act otherwise.

This tendency to continue in its state of rest or uniform motion is called inertia.

Several other tricks are based on the same principle. For example, if a pile of coins is placed on a table, the bottom one can be removed without disturbing the remainder simply by flicking it sharply with a piece of thin wood or metal. In both cases, the incidence of friction is insufficient to overcome inertia.

Sand on the beach

Before you step on it, the sand is packed as tightly as it can be under natural conditions. Your weight disturbs the sand, making the grains less efficiently packed. The sand is forced to occupy more volume and rises above the water level, becoming dry and white. The water rises more slowly, by capillary action, so it takes a few seconds or more before the sand gets wet and dark again.

Whats in the box?

The first box contains uncooked rice piled at an angle. The second box contains an iron weight and a pile of ice. It has to be finely balanced, so that the weight of the ice keeps the box on the table. In due course the ice will melt, toppling the box.

Quito

She would weigh zero. Her centre of gravity would be in orbit, considering the Earth's speed of rotation and the fact that Quito is on the Equator.

How do you get an egg into a bottle?

If the unpeeled egg is soaked overnight in vinegar, or acetic acid, its shell becomes plastic. The procedure described will put the egg in the bottle, and a cold-water rinse will restore the shell to its original hardness.

How does the Moon cause tides?

The Earth's rotation is gradually slowing down. Don't look forward to a 25-hour day, or to getting a few extra minutes of sleep in the morning, however. The effect is just enough to add about one second to the Earth's day every 100,000 years.

How do you photograph the Moon?

Photographs show that the Moon is the same size in both positions. The illusion is universal – it is apparent even in a planetarium – but there is still no fully accepted explanation for it. Ptolemy argued that the horizon Moon appears larger because we can compare it with distant trees and buildings. This theory is still the most widely accepted one, but it doesn't explain why sailors see the Moon illusion just as vividly at sea.

How long to travel through space?

It would seem that we can get there in a minimum of 4.3 years, but we must instead determine the maximum acceleration and deceleration a human body can stand to reach and return from a speed which has been approximated at 186,000 miles per second.

What's in the middle of a rainbow?

At the circle's centre is the shadow of your head. The droplets in a rainbow are on the surface of an imaginary cone that points straight to the Sun, behind you, and has its vertex at your eyes. (In the case of nearby rainbows, as in a lawn sprinkler, you may be able to see twin, overlapping bows, one from each eye.) A rainbow's arch is always 42 degrees away from the line of sight between you and your shadow. The full circle fills a visual angle of 84 degrees, nearly a right angle.

Climbing the mountain

Bob was most correct when he stated that the cold was due to the rarefied atmosphere. While it is true that the Sun's rays are stronger at greater heights and that winds have a cooling effect, and the greater the height the stronger the winds, the main reason for the lower temperature at greater heights is the rarefied atmosphere. With fewer molecules per cubic inch, fewer molecules bombard us each second and less heat is transmitted to us by the surrounding body of air.

The flawed sense

The sense we are referring to is sight.

Motion pictures consist of a succession of stills, usually 24 frames per second. The eye perceives these stills as smoothly flowing movements because of a phenomenon called 'persistence of vision.' The impression of an image on the retina lingers on, as it is retained by the brain for a fraction of a second after the image has disappeared.

How do you skim stones?

The first skip is short and to the right. When the stone's trailing edge hits, it pushes sand to the left; the stone tilts forward and hops to the right. Then the leading edge strikes and pushes sand to the right; the stone tilts back and takes a long hop to the left, and the cycle repeats. The short hops appear to be missing when stones skip over water. After the trailing edge strikes, the stone planes along, building a crest of water in front of it, then lifts out and makes a long hop. It strikes with its trailing edge again and repeats the process.

Rays of the sun

The Sun's rays are parallel, of course. The fanning out is an optical illusion, the same illusion that makes railroad tracks appear to meet at a point on the horizon. These rays are a more powerful illusion because of the absence of reference points. It's hard to believe, even when you know it for a fact, that two rays high overhead and so far apart are just as far apart in the distance where they first emerge from behind the cloud.

The space station

1. If the astronaut walked round the station in the direction of its rotation his weight would increase. If he turned round and walked in the opposite direction his weight would decrease. Why should this be? We know that the centripetal force increases

proportionally with the speed of rotation. By walking in that direction the astronaut adds his speed to that of the rotation.

2. Try to spin one of your coins on the floor of your room: the coin will refuse to spin. By conservation of angular momentum, a spinning object tries to maintain its position in space. Since the spinning station is continually changing your position in space, a coin that is spun will keep changing its orientation, to correct its angular momentum, and will topple and fall.

An even playing field?

According to Euclid, a plane can touch tangentially a sphere only at one point, X, as illustrated. Only one person standing at X can be considered upright with respect to the centre of the Earth. This is somewhat theoretical as the football pitch is negligibly small compared to the surface of the Earth. Nevertheless, it is perfectly true.

How do you tell acid from alkali?

A red rose petal. Press it hard against some paper, and the resultant pink stain will have the same properties as litmus. In fact a rose blossom can be made a deep blue with ammonia.

How do you build a bridge of bricks?

1. The arched design can accommodate a heavier burden.
2. The amount of material needed is the same for both designs.
3. If you slice away the humped portion of the arched bridge and place it in the gap underneath, you'll end up with a bridge identical to the straight one. This proves that there is the same amount of material in both.

Whisky and soda

There is the same amount of whisky in the soda as there is soda in the whisky. This classic puzzle can be attacked with algebra, fractions, and formulas, or with intuition.

Both containers are left with a quart of liquid after the transfer, the same volume they had to start with. The amount of whisky missing from its bottle is exactly matched by enough soda to bring the bottle's volume back up to a quart. Likewise, the amount of soda removed from its carton, whatever it may be, is now taken up by an equal volume of whisky. This is true of many pairs of liquids, but not with all – such as if a spoonful of one liquid plus a spoonful of the other becomes less (or more) than two spoonsful of mixture due to chemical reaction.

Can you see 3D?

The phenomenon of 3-D vision rests on the fact that the two eyes perceive slightly different versions of the same object.

The human brain is genetically programmed to integrate the two views of the same objects into three-dimensional vision. If you close one eye, the three-dimensional world you see around you becomes two-dimensional. You might not see much difference, but that is because your visual memory knows that what you are looking at is, in fact, three-dimensional and persists in seeing it that way. A one-eyed person, however, is unable to see in three dimensions so cannot estimate distances by parallax, an ability that relies on the processing of different images received by the two eyes.

Any attempt to make a two-dimensional image appear three-dimensional relies on deceiving the brain into reacting as if each eye were receiving a different image of the same object.

Years ago, the stereoscope, an optical instrument with a binocular eyepiece, presented two slightly different pictures or photographs for viewing, one with each eye. The brain was thus deceived and offered the illusion of a three-dimensional picture. The short-lived 3-D movie which presented overlapping images in red and green filtered out alternate viewpoints with the red- and green-lensed spectacles, again fooling the brain. These 3-D illusions, and some others, are based on the fact that human eyes usually focus on whatever object they are looking at. However, the eyes can defocus, or diverge, and this is the principle behind stereograms.

Let's do an experiment. Hold a pencil vertically right in front of your nose, relax, and then slowly move the pencil about fifteen inches away from your nose. You will most likely see two pencils, unless you try deliberately to focus. Now follow through with two further experiments which will lead to a full understanding of the stereogram phenomenon.

First, close one eye and then the other, alternatively. You will find that first one and then the other pencil will disappear. Second, turn the pencil to a horizontal position. The two pencils will merge into one, except for the two ends.

In a stereogram the patterns are duplicated, but with minute differences which are not observable because of the background. If you let your eyes wander, i.e., diverge or defocus, each eye will pick up a different version of the duplicated pattern. That is all the brain needs to produce 3-D vision. You can prove this theory with any stereogram pattern. Once you see the image in three dimensions, close one eye and the 3-D illusion will disappear. The same will happen if you turn the image sideways right or left, but it will reappear if you turn it upside down (180 degrees).

The wheelbarrow

Pulling a wheelbarrow is easier. Pushing presses the wheel harder against the ground, adding to the workload.

Air Mail?

In theory the scheme would work, although I doubt that any post office would pay you if they had to deliver a floating parcel.

It is commonly stated that spring balances measure weight (i.e., the force of gravity) and that beam balances measure mass (i.e., the amount of substance in an object). This is not strictly true; the beam balance compares weights, and therefore would not be affected by a change in gravity. For example, kitchen scales would give the same reading on the Moon, but a spring balance would not.

In this case, however, even a beam balance would not give a correct reading of the mass of the parcel. I do not think that there is a common instrument that would. Either the parcel should be weighed in a vacuum or a machine would have to be invented that measured the inertial mass by, perhaps, gently oscillating the parcel.

The Spying Game

No, the operation would not work theoretically. A minute variation in temperature would completely demolish the code, due to thermal expansion.

You could cope with the problem of thermal expansion by converting the decimal into a fraction and cutting the bar into two pieces, one to represent the numerator and the other the denominator. This would be purely a mathematical solution, inasmuch as thermal expansion affects both bars, leaving the ratio, and therefore the fraction, unaffected.

On a physical level it would not even work theoretically. The problem is the tremendously small distances that one is soon involved in. Every letter added to the message would involve a hundredfold increase in accuracy. Sending messages with only two letters (e.g., OK) would probably be feasible, for this would require an accuracy of 0.1mm.

To send a five-letter message (e.g., ATOMS) would require an accuracy of 0.1 nanometres, which is about half the diameter of an atom of iron. As you cannot have half an atom, the sending of such a message by this method is impossible.

You might think that lengthening the bar would provide a viable solution. Ignoring the logistic problem connected with sending a long bar, let us re-examine the position. To send the message ISOTOPE MASS, we have to transmit the number 0919152015160539130111919. Let us suppose we send a bar whose length is this number of millimeters. How long is the bar?

91,915,201,516,053,913,011,919 millimeters

$= 10^{23}$ mm

$= 10^{17}$ km, 600 million times farther than the distance (93 million miles) between the Earth and the Sun.

Note that it may be possible to work out a method using separate bars for each letter, possibly with the length in cm indicating letter order, and tenths of mm as the code.

How heavy can you get?

Yes and no. Our weight is conveyed by the gravitational attraction of the Earth. However, part of this force is counterbalanced by the centripetal force due to the rotation of the globe. This force is zero at the poles, increasing to a maximum at the Equator. A person standing, for example, in Quito will have his centre of gravity move up as he grows, thereby increasing the centripetal force, while at the same time reducing the effect of gravity. Consequently, a point will come when the resultant weight will start to decrease. When the centre of gravity is about 22,000 miles (36,000km) high, the person would be weightless and efficiently in orbit. Needless to say, at one of the poles, the weight would keep on increasing, irrespective of height, until the person's mass was enough to exert measurable gravity on the earth.

Can you float a battleship in a bathtub?

Strange as it seems, it is perfectly possible to float a battleship in a bathtub. As long as there is enough water to surround the ship completely, it will float. The ship's hull can't 'tell' whether it is surrounded by an expanse of ocean or by a mere fraction of an inch of water. The water pressure on the hull is the same in either case. Likewise, the hydrostatic pressure of the ship is independent of the amount of water that is below or to the side of the ship. Many people find this hard to believe because they confuse the amount of water displaced with the amount necessary to float the ship. Look at it this way: Suppose the ship weighs 30,000 tons. If the tub is filled to the rim with several million gallons of water and you lower the ship in, the amount of water that will spill over the sides will, indeed, weigh 30,000 tons. But that's water over the rim. The amount of water left inside, floating the ship, could be a great deal less. If the 'fit' between the ship's hull and the tub's wall is tight enough, the water could be squeezed into a thin envelope only an inch thick or less, completely surrounding the ship. This principle is put into practice at Mount Palomar Observatory, where the giant 530-ton horseshoe telescope actually floats in a basin on a thin cushion of oil.

The Light and the Shadow

The top of the shadow moves faster than the man.

As proof, let A be the position of the man at one point in time, and B be the man's position after he has walked 20 metres towards the light. Let AA and BB be the top of the shadows in the two positions. The distance AA-BB is clearly greater than A-B so, consequently, the top of the shadow must have moved faster.

Feathers and Gold I

A pound of feathers weighs more. This is because feathers are weighed using the Avoirdupois system, whereas gold (as are silver and drugs) is weighed using the Troy system. A pound on the Avoirdupois scale is greater than a Troy pound.

Glossary

Assuming that readers may not be familiar with all the scientific terms used in this book, it should be useful, for ease of reference, to define here some concepts not part of everyday vocabulary

absolute zero – the temperature at which the molecules of any object are stationary; this, therefore, represents the coldest possible temperature.

Adhesion/cohesion – adhesion is the force of attraction between two different types of molecule. The force of attraction between similar molecules is called cohesion.

Adiabatic process – a term used in thermodynamics, referring to a condition in which no heat enters or leaves the system, although pressure and volume are varied. For example, the air in a bicycle pump will heat up if compressed, because no significant heat transfer will take place immediately. Another example, with the reverse effect, is an aerosol can. On releasing the contents, the temperature of the can will drop. Other systems in daily use, such as automobile engines and refrigerators, exhibit adiabatic phenomena.

Archimedes' principle – states that when any object is immersed in a fluid there is an upthrust that acts on the object which equals the weight of the displaced fluid.

Bernoulli effect – discovered and formulated by the Swiss mathematician Daniel Bernoulli in 1738. The principle states that, as the speed of a moving liquid or gas increases, the pressure within that fluid decreases. This principle is an important aspect of aerodynamics and covers the flow over surfaces, such as the wings of aircraft and ship's propellers. As the air flows over the upper surface of a wing it speeds up, and consequently suffers a reduction in pressure as compared to the lower surface. The resulting difference in pressure provides lift to the aircraft.

Centre of gravity – the point in any object where all the weight could be considered to be concentrated without affecting the properties of the object. Such an assumption can greatly simplify calculations.

cohesion – *See* **adhesion**.

conduction of heat – the process by which heat energy can travel through a substance by the transfer of vibrational energy between adjacent molecules. This is greatest in solids.

convection of heat – the process by which hot fluid is forced to rise by being displaced by colder fluid.

Coriolis effect – a deflection caused by the rotation of the Earth.

diffraction – a process by which a wave curls around small objects or spreads out as a result of passing through a small gap.

drag – *See* **friction**.

energy conservation – the scientific law that states that energy (heat, light, sound, electricity, etc.) cannot be created or destroyed, but can only be converted into another form.

evolution – the theory proposed by Charles Darwin that states that species change over long periods of time, as a result of which individuals that possess advantageous differences are more successful in procreation.

fluid – a substance that flows: a liquid or a gas.

friction – a force that results from one object moving relative to another with which it is in contact.

gravity – a force of attraction between any two objects. The force is very small and is only noticeable if at least one of the objects is very large.

LED – stands for Light Emitting Diode, a small solid-state device that produces light, without heat, when a small electrical current is passed through it.

lift – the upward force produced by the flow of air over wings. If the lift disappears because of a disruption in this flow, the wing is said to stall.

light-year – the distance that light travels in a vacuum in one year, about 5.88 trillion miles (or 9,460,000,000,000,000 metres).

mass – measure of how much material there is in an object. This is independent of the force of gravity (weight) acting on the object. Weight is measured by a spring balance and mass by a beam balance.

Newton's laws of motion – (1) An object will remain at rest or in uniform motion unless acted upon by an unbalanced force. (2) When an unbalanced force acts upon an object, it will accelerate at a rate proportional to the force and inversely proportional to the mass of the object. (3) For any action on an object, there will be an equal and opposite reaction.

perpetual motion machine – refers to a mechanical device doing work and operating perpetually without any supply of energy, other than that which is generated by the device itself. Such a system cannot exist, as it is contrary to a well-established physical law, namely the principle of conservation of energy. This law is so fundamental that patent offices have been known to refuse any application based on perpetual motion.

Perpetual motion is perfectly feasible in the absence of friction; for example, electrons rotating around a nucleus or an object flying through space. However, as soon as energy is extracted from such a system, it will slow down, destroying its perpetual motion. Needless to say, any device using the force of gravity or variation of atmospheric pressure or temperature would not qualify.

polarization of light – light is a waveform that has an electric and magnetic vector. In all forms of natural light these vectors are not aligned. If aligned, the light is said to be polarized.

polarizing filter – A filter that allows light to pass which is vibrating in one direction only. See polarization of light.

principle of flotation – states that a floating object displaces its own weight of fluid.

radiation of heat – the process by which heat can travel through a vacuum or a fluid. It is a form of light (infrared).

Rayleigh scattering – named after Lord Rayleigh (1842-1919), a type of deflection of electromagnetic radiation by particles in the matter through which it passes. The radiation photons bounce off the atoms and molecules without any change of energy (elastic scattering), changing phase but not frequency, as opposed to inelastic scattering.

refraction – the process by which the direction of light changes when it travels from one medium to another of different density. This is how lenses work.

stall – *See* **lift.**

surface tension – a condition existing on the surface of liquids giving them film-like characteristics. The tension is explained as resulting from intermolecular forces of the liquid. Examples: A water beetle can ride the surface of water; the near-perfect sphere of a soap bubble; a small quantity of mercury poured onto a plane assumes a near-spherical shape, flattened only slightly by gravity.

weight – *See* **mass.**